BASIC CONCEPTS IN | RELATIVITY
| and
| EARLY QUANTUM THEORY

BASIC CONCEPTS IN

NEW YORK · LONDON · SYDNEY · TORONTO

RELATIVITY
and
EARLY QUANTUM THEORY

ROBERT RESNICK

Professor of Physics
Rensselaer Polytechnic Institute

John Wiley and Sons, Inc.

Library of Congress Cataloging in Publication Data

Resnick, Robert, 1923–
    Basic concepts in relativity and early quantum theory.

    Includes bibliographical references.
    1. Relativity (Physics)   2. Quantum theory.
I. Title
QC6.R3877   530.1′1   78-39835

ISBN 0-471-71702-9
ISBN 0-471-71703-7 (pbk.)

Printed in the United States of America.

10 9 8 7 6 5 4 3 2 1

To My Mother

Preface

Relativity and quantum theory form the conceptual basis for what is called modern physics. During the past decade, modern physics has come to occupy an increasing portion of introductory physics courses. This development has been followed, more recently, by a widespread reduction in the length of these courses, which has forced a reassessment of course structure and content. This book is one outcome of such a reassessment.

There has been a growing interest in the teaching of special relativity in general physics courses. Many new paperbacks, including one of my own,* were written to satisfy this interest. But, despite much student enthusiasm for relativity, the reduction in course lengths created a need for a somewhat briefer and necessarily less sophisticated treatment than was permissible before. Another trend that required adjustment was the explicit introduction of quantum mechanics and its applications into introductory courses. Often this material required an entire semester, such as the last semester of a four-semester course. Even at schools that have such two-year courses, however, many students were terminating their study after three semesters and, at the majority of schools that have shorter courses, the material appeared first in a junior-level modern physics course. What many called for was a new treatment of quantum concepts in general physics that would lend coherence to these shorter courses and set the stage for students who go on to a new course.

As a result, I have written *Basic Concepts in Relativity and Early Quantum Theory*. It can be regarded as an alternative ending to the text, *Physics,*** and—according to colleagues who have persuaded me to do it—may serve other useful teaching purposes as well. Thus the treatment of relativity logically follows the optics section of *Physics* and is a briefer and simpler treatment than was given in the previously mentioned paperback. Similarly, the treatment of early quantum theory is an expanded and a conceptually deeper one than is presently afforded by the last two chapters of *Physics*. This work matches in every way the level, style, and other special characteristics of

---

*Robert Resnick, *Introduction to Special Relativity,* Wiley, New York, 1968.
**David Halliday and Robert Resnick, *Physics,* 2nd edition, Wiley, New York, 1966.

vii

Physics in two areas—relativity and early quantum theory—which are now regarded as essential to shorter introductory courses. It provides a logical and modern ending to that course and a foundation for subsequent extensions of quantum physics.

There are many different ways to use this material. For example, a significant part of it can be regarded as optional. Hence there are sections within chapters—clearly labeled as optional—on special, historical, or more advanced topics, and some whole chapters (such as Chapter 4), that can be skipped without a serious lack of continuity. Therefore, with such deletions, this book presents an alternative ending to *Fundamentals of Physics** (a shorter and less sophisticated version of *Physics*) as well. Various coherent uses of the material could require from as little as four weeks to perhaps as much as ten weeks of time in typical courses. In addition, the book could prove useful on its own in a variety of settings, such as in summer courses for teachers and as a basis for minicourses that are now becoming fashionable in this era of calendar reform. Many other uses will suggest themselves to experienced professors.

In each chapter, references are cited to encourage students to read original or popular sources. Often, in brackets, a reference is given to earlier sections of *Physics,* which can serve as a review of the classical physics assumed known. A set of more than 350 questions and problems, spanning a wide range of content and level of difficulty, is provided to enable instructors to vary the emphasis—the most difficult problems being starred. Optional sections— which often have an intrinsic interest of their own—are indicated by a gray bar at the side. The writing is expansive and pedagogic aids, such as summary tables and worked-out examples, are employed to help the student to learn on his own. Answers to problems and tables of useful data and relations are also provided.

I am grateful to many persons who assisted me in the preparation of this book, but especially to Stephen Arendt, Benjamin Chi, Carolyn Clemente, Donald Deneck, Phyllis Kallenburg, and Paul Stoler. I hope that this work will contribute to the improvement of physics education.

Robert Resnick

Troy, New York
January, 1972

*David Halliday and Robert Resnick, *Fundamentals of Physics,* Wiley, New York, 1970.

Contents

Introduction to Chapters 1 to 3

Modern physics can be defined as physics that requires relativity theory or quantum theory for its interpretation. These theories emerged in the early decades of the twentieth century as the classical theories encountered increasing difficulty in explaining experimental observations.

In this first section of the book (Chapters 1 to 3), we examine the experimental background to relativity, the development of the special theory of relativity, and the experimental confirmation of relativistic predictions. We shall see that classical mechanics breaks down in the region of very high speeds and that relativistic mechanics is a generalization that includes the classical laws as a special case. Gradually the student will develop a physical feeling for the principles of relativity. The point of view and the results that emerge from relativistic considerations prove to be useful and necessary in many areas of modern physics, including atomic, nuclear, and high energy physics.

The Experimental Background
of the Theory of Special Relativity

1.1 Introduction

To send a signal through free space from one point to another as fast as possible, we use a beam of light or some other electromagnetic radiation such as a radio wave. *No faster method of signaling has ever been discovered.* This experimental fact suggests that the speed of light in free space, c $(= 3.00 \times 10^8$ m/sec),* is an appropriate limiting reference speed to which other speeds, such as the speeds of particles or of mechanical waves, can be compared.

In the macroscopic world of our ordinary experiences, the speed u of moving objects or mechanical waves with respect to any observer is always much less than c. For example, an artificial satellite circling the earth may move at 18,000 mph with respect to the earth; here $u/c = 0.000027$. Sound waves in air at room temperature move at 332 m/sec through the air so that $u/c = 0.0000010$. It is in this ever-present, but limited, macroscopic environment that our ideas about space and time are first formulated and in which Newton developed his system of mechanics.

In the microscopic world, however, it is possible to find particles whose speeds are quite close to that of light. For an electron accelerated through a 10-million-volt potential difference, a value reasonably easy to obtain, the speed u equals $0.9988c$. We cannot be certain without direct experimental test that Newtonian mechanics can be safely extrapolated from the ordinary region of low speeds $(u/c \ll 1)$ in which it was developed to this high-speed region $(u/c \rightarrow 1)$. Experiment shows, in fact, that Newtonian mechanics does *not* predict the correct answers when it is applied to such fast particles. Indeed, in Newtonian mechanics there is no limit in principle to the speed attainable by a particle, so that the speed of light c should play no special role at all.

---

*The presently accepted value of the speed of light is $2.997925 \pm 0.000003 \times 10^8$ m/sec.

And yet, if the energy of the 10 Mev electron above is increased by a factor of four (to 40 Mev) experiment [1] shows that the speed is not doubled to $1.9976c$, as we might expect from the Newtonian relation $K = \frac{1}{2}Mv^2$, but remains below c; it increases only from $0.9988c$ to $0.9999c$, a change of 0.11 percent. Or, if the 10 Mev electron moves at right angles to a magnetic field of 2.0 tesla, the measured radius of curvature of its path is not 0.53 cm (as may be computed from the classical relation $r = m_e v/qB$) but, instead, 1.8 cm. Hence, no matter how well Newtonian mechanics may work at low speeds, it fails badly as $u/c \rightarrow 1$.

In 1905 Albert Einstein published his special theory of relativity. Although motivated by a desire to gain deeper insight into the nature of electromagnetism, Einstein, in his theory, extended and generalized Newtonian mechanics as well. He correctly predicted the results of mechanical experiments over the complete range of speeds from $u/c = 0$ to $u/c \rightarrow 1$. Newtonian mechanics was revealed to be an important special case of a more general theory. In developing this theory of relativity, Einstein critically examined the procedures used to measure length and time intervals. These procedures require the use of light signals and, in fact, an assumption about the way light is propagated is one of the two central hypotheses upon which the theory is based. His theory resulted in a completely new view of the nature of space and time.

The connection between mechanics and electromagnetism is not surprising because light, which (as we shall see) plays a basic role in making the fundamental space and time measurements that underlie mechanics, is an electromagnetic phenomenon. However, our low-speed Newtonian environment is so much a part of our daily life that almost everyone has some conceptual difficulty in understanding Einstein's ideas of space-time when he first studies them. Einstein may have put his finger on the difficulty when he said "Common sense is that layer of prejudices laid down in the mind prior to the age of eighteen." Indeed, it has been said that every great theory begins as a heresy and ends as a prejudice. The ideas of motion of Galileo and Newton may very well have passed through such a history already. More than a half-century of experimentation and application has removed special relativity theory from the heresy stage and put it on a sound conceptual and practical basis. Furthermore, we shall show that a careful analysis of the basic assumptions of Einstein and of Newton makes it clear that the assumptions of Einstein are really much more reasonable than those of Newton.

In the following pages, we shall develop the experimental basis for the ideas of special relativity theory. Because, in retrospect, we found that Newtonian mechanics fails when applied to high-speed particles, it seems wise to begin by examining the foundations of Newtonian mechanics. Perhaps, in this way, we can find clues as to how it might be generalized to yield correct results at high speeds while still maintaining its excellent agreement with experiment at low speeds.

1.2 Galilean Transformations

Let us begin by considering a physical *event*. An event is something that happens independently of the reference frame we might use to describe it.

For concreteness, we can imagine the event to be a collision of two particles or the turning-on of a tiny light source. The event happens at a point in space and at an instant in time. We specify an event by four (space-time) measurements in a particular frame of reference, say the position numbers x, y, z and the time t. For example, the collision of two particles may occur at $x = 1$ m, $y = 4$ m, $z = 11$ m, and at time $t = 7$ sec in one frame of reference (e.g., a laboratory on earth) so that the four numbers $(1, 4, 11, 7)$ specify the event in that reference frame. The same event observed from a different reference frame (e.g., an airplane flying overhead) would also be specified by four numbers, although the numbers may be different than those in the laboratory frame. Thus, if we are to describe events, our first step is to establish a frame of reference.

We define an *inertial system* as a frame of reference in which the law of inertia—Newton's first law—holds. In such a system, which we may also describe as an *unaccelerated* system, a body that is acted on by zero net external force will move with a constant velocity. Newton assumed that a frame of reference fixed with respect to the stars is an inertial system. A rocket ship drifting in outer space, without spinning and with its engines cut off, provides an ideal inertial system. Frames accelerating with respect to such a system are not inertial.

In practice, we can often neglect the small (acceleration) effects due to the rotation and the orbital motion of the earth and to solar motion.* Thus, we may regard any set of axes fixed on the earth as forming (approximately) an inertial coordinate system. Likewise, any set of axes moving at uniform velocity with respect to the earth, as in a train, ship, or airplane, will be (nearly) inertial because motion at uniform velocity does not introduce acceleration. However, a system of axes which accelerates with respect to the earth, such as one fixed to a spinning merry-go-round or to an accelerating car, is *not* an inertial system. A particle acted on by zero net external force will not move in a straight line with constant speed according to an observer in such noninertial systems.

The special theory of relativity, which we consider here, deals only with the description of events by observers in inertial reference frames. The objects whose motions we study may be accelerating with respect to such frames but the frames themselves are unaccelerated. The general theory of relativity, presented by Einstein in 1917, concerns itself with all frames of reference, including noninertial ones.**

Consider now an inertial frame S and another inertial frame S' which moves at a constant velocity $\mathbf{v}$ with respect to S, as shown in Fig. 1-1. For convenience, we choose the three sets of axes to be parallel and allow their relative motion to be along the common x, x' axis. We can easily generalize to arbitrary orientations and relative velocity of the frames later, but the physical

---

*Situations in which these effects are noticeable are the Foucault pendulum experiment or the deflection from the vertical of a freely falling body. The order of magnitude of such effect is indicated by the result that in falling vertically 100 ft (1200 in.) a body at the Equator is deflected less than $\frac{1}{6}$ in. from the vertical.

**See *Introduction to Special Relativity*, Robert Resnick, (John Wiley and Sons, Inc., New York, 1968), Supplementary Topic C, for a brief discussion of general relativity theory.

principles involved are not affected by the particular simple choice we make at present. Note also that we can just as well regard S to be moving with velocity $-\mathbf{v}$ with respect to S' as we can regard S' to move with velocity $\mathbf{v}$ with respect to S.

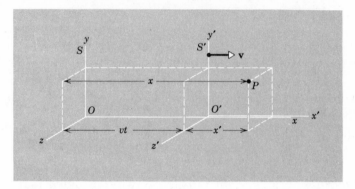

Figure 1-1. Two inertial frames with a common x-x' axis and with the y-y' and z-z' axes parallel. As seen from frame S, frame S' is moving in the positive x-direction at speed v. Similarly, as seen from frame S', frame S is moving in the negative x'-direction at this same speed. Point P suggests an *event*, whose space-time coordinates may be measured by each observer. The origins O and O' coincide at time $t = 0$, $t' = 0$.

Let an event occur at point P, whose space and time coordinates are measured in each inertial frame. An observer attached to S specifies by means of meter sticks and clocks, for instance, the location and time of occurrence of this event, ascribing space coordinates x, y, and z and time t to it. An observer attached to S', using his measuring instruments, specifies the *same* event by space-time coordinates x', y', z' and t'. The coordinates x, y, z will give the position of P relative to the origin O as measured by observer S, and t will be the time of occurrence of P that observer S records with his clocks. The coordinates x', y', and z' likewise refer the position of P to the origin O' and the time of P, t', to the clocks of inertial observer S'.

We now ask what the relationship is between the measurements $x, y, z,$ t and x', y', z', t'. The two inertial observers use meter sticks, which have been compared and calibrated against one another, and clocks, which have been synchronized and calibrated against one another. The classical procedure, which we look at more critically later, is to assume thereby that length intervals and time intervals are absolute, that is, that they are the same for all inertial observers of the same events. For example, if meter sticks are of the same length when compared at rest with respect to one another, it is implicitly assumed that they are of the same length when compared in relative motion to one another. Similarly, if clocks are calibrated and synchronized when at rest, it is assumed that their readings and rates will agree thereafter, even if they are put in relative motion with respect to one another. These are examples of the "common sense" assumptions of classical theory.

We can state these results explicitly, as follows. For simplicity, let us say that the clocks of each observer read zero at the instant that the origins O

and O' of the frames S and S', which are in relative motion, coincide. Then the *Galilean coordinate transformations*, which relate the measurements x, y, z, t to x', y', z', t', are

$$\begin{aligned} x' &= x - vt \\ y' &= y \\ z' &= z. \end{aligned} \tag{1-1a}$$

These equations agree with our classical intuition, the basis of which is easily seen from Fig. 1-1. It is assumed that time can be defined independently of any particular frame of reference. This is an implicit assumption of classical physics, which is expressed in the transformation equations by the absence of a transformation for t. We can make this assumption of the universal nature of time explicit by adding to the Galilean transformations the equation

$$t' = t. \tag{1-1b}$$

It follows at once from Eqs. 1-1a and 1-1b that the time interval between occurrence of two given events, say P and Q, is the same for each observer, that is

$$t'_P - t'_Q = t_P - t_Q, \tag{1-2a}$$

and that the distance, or space interval, between two points, say A and B, measured at a given instant, is the same for each observer, that is

$$x'_B - x'_A = x_B - x_A. \tag{1-2b}$$

♦ **Example 1.** Derive the classical space-interval result, Eq. 1-2b.

Let A and B be the end points of a rod, for example, which is at rest in the S-frame. Then, the primed observer, for whom the rod is moving with velocity $-\mathbf{v}$, will measure the end-point locations as x'_B and x'_A, whereas the unprimed observer locates them at x_B and x_A. Using the Galilean transformations, however, we find that

$$x'_B = x_B - vt_B \qquad \text{and} \qquad x'_A = x_A - vt_A,$$

so that

$$x'_B - x'_A = x_B - x_A - v(t_B - t_A).$$

Since the two end points, A and B, are measured at the same instant, we must put $t_A = t_B$ and we obtain

$$x'_B - x'_A = x_B - x_A,$$

as found above.

Or, we can imagine the rod to be at rest in the primed frame, and moving therefore with velocity $\mathbf{v}$ with respect to the unprimed observer. Then the Galilean transformations, which can be written equivalently as

$$\begin{aligned} x &= x' + vt \\ y &= y' \\ z &= z' \\ t &= t', \end{aligned} \tag{1-3}$$

give us $x_B = x'_B + vt'_B$ and $x_A = x'_A + vt'_A$ and, with $t'_A = t'_B$, we once again obtain $x_B - x_A = x'_B - x'_A$.

Notice carefully that two measurements (the end points x'_A, x'_B or x_A, x_B) are made for each observer and that we assumed they were made at the *same time* ($t_A = t_B$, or

$t'_A = t'_B$). The assumption that the measurements are made at the same time—that is, simultaneously—is a crucial part of our definition of the length of the moving rod. Surely we should not measure the locations of the end points at different times to get the length of the moving rod; it would be like measuring the location of the tail of a swimming fish at one instant and of its head at another instant in order to determine its length (see Fig. 1-2). ◀

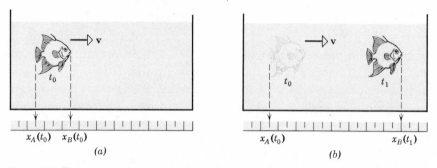

Figure 1-2. To measure the length of a swimming fish, one must mark the positions of its head and tail simultaneously (*a*), rather than at arbitrary times (*b*).

The time-interval and space-interval measurements made above are absolutes according to the Galilean transformation; that is, they are the same for all inertial observers, the relative velocity **v** of the frames being arbitrary and not entering into the results. When we add to this result the assumption of classical physics that the mass of a body is a constant, independent of its motion with respect to an observer, then we can conclude that classical mechanics and the Galilean transformations imply that length, mass, and time—the three basic quantities in mechanics—are all independent of the relative motion of the measurer (or observer).

1.3 Newtonian Relativity

How do the measurements of different inertial observers compare with regard to *velocities* and *accelerations* of objects? The position of a particle in motion is a function of time, so that we can express particle velocity and acceleration in terms of time derivatives of position. We need only carry out successive time differentiations of the Galilean transformations. The *velocity* transformation follows at once. Starting from

$$x' = x - vt,$$

differentiation with respect to t gives

$$\frac{dx'}{dt} = \frac{dx}{dt} - v.$$

But, because $t = t'$, the operation d/dt is identical to the operation d/dt', so that

$$\frac{dx'}{dt} = \frac{dx'}{dt'}.$$

Therefore,

$$\frac{dx'}{dt'} = \frac{dx}{dt} - v.$$

Similarly,

$$\frac{dy'}{dt'} = \frac{dy}{dt}$$

and

$$\frac{dz'}{dt'} = \frac{dz}{dt}.$$

However, $dx'/dt' = u'_x$, the x-component of the velocity measured in S', and $dx/dt = u_x$, the x-component of the velocity measured in S, and so on, so that we have simply the *classical velocity addition theorem*

$$\begin{aligned} u'_x &= u_x - v \\ u'_y &= u_y \\ u'_z &= u_z. \end{aligned} \tag{1-4}$$

Clearly, in the more general case in which **v**, the relative velocity of the frames, has components along all three axes, we would obtain the more general (vector) result

$$\mathbf{u'} = \mathbf{u} - \mathbf{v}. \tag{1-5}$$

The student has already encountered many examples of this (see Resnick and Halliday, *Physics,* Part I, Sec. 4-6). For example, the velocity of an airplane with respect to the air (**u'**) equals the velocity of the plane with respect to the ground (**u**) minus the velocity of the air with respect to the ground (**v**).

▶ **Example 2.** A passenger walks forward along the aisle of a train at a speed of 2.2 mi/hr as the train moves along a straight track at a constant speed of 57.5 mi/hr with respect to the ground. What is the passenger's speed with respect to the ground?

Let us choose the train to be the primed frame so that $u'_x = 2.2$ mi/hr. The primed frame moves forward with respect to the ground (unprimed frame) at a speed $v = 57.5$ mi/hr. Hence, the passenger's speed with respect to ground is

$$u_x = u'_x + v = 2.2 \text{ mi/hr} + 57.5 \text{ mi/hr} = 59.7 \text{ mi/hr}.$$

Example 3. Two electrons are ejected in opposite directions from radioactive atoms in a sample of radioactive material at rest in the laboratory. Each electron has a speed $0.67c$ as measured by a laboratory observer. What is the speed of one electron as measured from the other, according to the classical velocity addition theorem?

Here, we may regard one electron as the S frame, the laboratory as the S' frame, and the other electron as the object whose speed in the S-frame is sought (see Fig. 1-3). In the S'-frame, the other electron's speed is $0.67c$, moving in the positive x'-direction say, and the speed of the S-frame (one electron) is $0.67c$, moving in the negative x'-direction. Thus, $u'_x = +0.67c$ and $v = +0.67c$, so that the other electron's speed with respect to the S-frame is

$$u_x = u'_x + v = +0.67c + 0.67c = +1.34c,$$

according to the classical velocity addition theorem. ◀

To obtain the acceleration transformation we merely differentiate the velocity relations (Eq. 1-4). Proceeding as before, we obtain

$$\frac{d}{dt'}(u'_x) = \frac{d}{dt}(u_x - v),$$

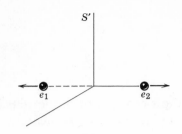

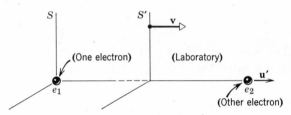

Figure 1-3. (*a*) In the laboratory frame, the electrons are observed to move in opposite directions at the same speed. (*b*) In the rest frame, *S*, of one electron, the laboratory moves at a velocity **v.** In the laboratory frame, *S'*, the second electron has a velocity denoted by **u'.** What is the velocity of this second electron as seen by the first?

or
$$\frac{du'_x}{dt'} = \frac{du_x}{dt},\qquad v \text{ being a constant,}$$

$$\frac{du'_y}{dt'} = \frac{du_y}{dt},$$

and
$$\frac{du'_z}{dt'} = \frac{du_z}{dt}.$$

That is, $a'_x = a_x$, $a'_y = a_y$, and $a'_z = a_z$. Hence, $\mathbf{a'} = \mathbf{a}$. The measured components of acceleration of a particle are unaffected by the uniform relative velocity of the reference frames. The same result follows directly from two successive differentiations of Eqs. 1-1 and applies generally when **v** has an arbitrary direction, as long as **v** = constant.

We have seen that different velocities are assigned to a particle by different observers when the observers are in relative motion. These velocities always *differ by* the relative velocity of the two observers, which in the case of inertial observers is *a constant velocity*. It follows then that when the particle velocity changes, the *change will be the same* for both observers. Thus, they each measure the *same acceleration* for the particle. The acceleration of a particle is the same in *all* reference frames which move relative to one another with constant velocity; that is

$$\mathbf{a'} = \mathbf{a}. \tag{1-6}$$

In classical physics the *mass* is also unaffected by the motion of the reference frame. Hence, the product *m***a** will be the same for all inertial observers. If

$\mathbf{F} = m\mathbf{a}$ is taken as the definition of force, then obviously each observer obtains the same measure for each force. With $\mathbf{F} = m\mathbf{a}$ and $\mathbf{F}' = m\mathbf{a}'$ it follows from Eq. 1-6 that $\mathbf{F} = \mathbf{F}'$. *Newton's laws of motion and the equations of motion of a particle would be exactly the same in all inertial systems.* Since, in mechanics, the conservation principles—such as those for energy, linear momentum, and angular momentum—all can be shown to be consequences of Newton's laws, it follows that *the laws of mechanics are the same in all inertial frames.* Let us make sure that we understand just what this paragraph says before we draw some important conclusions from it.

Although different inertial observers will record different velocities for the same particle, and hence different momenta and kinetic energies, they will agree that momentum is conserved in a collision or is not conserved, that mechanical energy is conserved or is not conserved, and so forth. The tennis ball on the court of a moving ocean liner will have a different velocity to a passenger than it has for an observer on shore, and the billiard balls on the table in a home will have different velocities to the player than they have for an observer on a passing train. But, whatever the values of the particle's or system's momentum or mechanical energy may be, when one observer finds that they do not change in an interaction, the other observer will find the same thing. Although the numbers assigned to such things as velocity, momentum, and kinetic energy may be different for different inertial observers, the laws of mechanics (e.g., Newton's laws and the conservation principles) will be the same in all inertial systems (see Problems 2 to 6). This is illustrated in the following example.

▶ **Example 4.** A particle of mass $m_1 = 3$ kg, moving at a velocity of $u_1 = +4$ m/sec along the x-axis of frame S, approaches a second particle of mass $m_2 = 1$ kg, moving at a velocity $u_2 = -3$ m/sec along this axis. After a head-on collision, it is found that m_2 has a velocity $U_2 = +3$ m/sec along the x-axis.

(a) Calculate the expected velocity U_1 of m_1, after the collision.

We use the law of conservation of momentum.

Before the collision the momentum of the system of two particles is

$$P = m_1u_1 + m_2u_2 = (3 \text{ kg})(+4 \text{ m/sec}) + (1 \text{ kg})(-3 \text{ m/sec})$$
$$= +9 \text{ kg-m/sec.}$$

After the collision the momentum of the system,

$$P = m_1U_1 + m_2U_2,$$

is also $+9$ kg-m/sec, so that

$$+9 \text{ kg-m/sec} = (3 \text{ kg})(U_1) + (1 \text{ kg})(+3 \text{ m/sec})$$

or
$$U_1 = +2 \text{ m/sec along the x-axis.}$$

(b) Discuss the collision as seen by observer S' who has a velocity $\mathbf{v}$ of $+2$ m/sec relative to S along the x-axis.

The four velocities measured by S' can be calculated from the Galilean velocity transformation equation (Eq. 1-5), $\mathbf{u}' = \mathbf{u} - \mathbf{v}$, from which we get

$$u_1' = u_1 - v = +4 \text{ m/sec} - 2 \text{ m/sec} = 2 \text{ m/sec,}$$
$$u_2' = u_2 - v = -3 \text{ m/sec} - 2 \text{ m/sec} = -5 \text{ m/sec,}$$
$$U_1' = U_1 - v = +2 \text{ m/sec} - 2 \text{ m/sec} = 0,$$
$$U_2' = U_2 - v = +3 \text{ m/sec} - 2 \text{ m/sec} = 1 \text{ m/sec.}$$

The system momentum in S' is

$$P' = m_1 u_1' + m_2 u_2' = (3 \text{ kg})(2 \text{ m/sec}) + (1 \text{ kg})(-5 \text{ m/sec})$$
$$= +1 \text{ kg-m/sec}$$

before the collision, and

$$P' = m_1 U_1' + m_2 U_2' = (3 \text{ kg})(0) + (1 \text{ kg})(1 \text{ m/sec})$$
$$= +1 \text{ kg-m/sec}$$

after the collision.

Hence, although the velocities and momenta have different numerical values in the two frames, S and S', when momentum is conserved in S it is also conserved in S'. ◀

An important consequence of the above discussion is that *no mechanical experiments carried out entirely in one inertial frame can tell the observer what the motion of that frame is with respect to any other inertial frame.* The billiard player in a closed box-car of a train moving uniformly along a straight track cannot tell from the behavior of the balls what the motion of the train is with respect to ground. The tennis player in an enclosed court on an ocean liner moving with uniform velocity (in a calm sea) cannot tell from his game what the motion of the boat is with respect to the water. No matter what the relative motion may be (perhaps none), so long as it is constant, the results will be identical. Of course, we *can* tell what the *relative* velocity of two frames may be by comparing the data the different observers take on the very same event—but then we have not deduced the relative velocity from observations *confined to a single frame.*

Furthermore, there is no way at all of determining the *absolute* velocity of an inertial reference frame from our mechanical experiments. No inertial frame is preferred over any other, for the laws of mechanics are the same in all. Hence, there is no physically definable absolute rest frame. We say that all inertial frames are equivalent as far as mechanics is concerned. The person riding the train cannot tell absolutely whether he alone is moving, or the earth alone is moving past him, or if some combination of motions is involved. Indeed, would you say that you on earth are at rest, that you are moving 30 km/sec (the speed of the earth in its orbit about the sun) or that your speed is much greater still (for instance, the sun's speed in its orbit about the galactic center)? Actually, no mechanical experiment can be performed which will detect an absolute velocity through empty space. This result, that we can only speak of the *relative* velocity of one frame with respect to another, and not of an absolute velocity of a frame, is sometimes called *Newtonian relativity.*

Transformation laws, in general, will change many quantities but will leave some others unchanged. These unchanged quantities are called invariants of the transformation. In the Galilean transformation laws for the relation between observations made in different inertial frames of reference, for example, acceleration is an invariant and—more important—so are Newton's laws of motion. A statement of what the invariant quantities are is called a relativity principle; it says that for such quantities the reference frames are equivalent to one another, no one having an absolute or privileged status relative to the others. Newton expressed his relativity principle as follows: "The motions of bodies included in a given space are the same amongst themselves, whether that space is at rest or moves uniformly forward in a straight line."

1.4 Electromagnetism and Newtonian Relativity

Let us now consider the situation from the electrodynamic point of view. That is, we inquire now whether the laws of physics other than those of mechanics (such as the laws of electromagnetism) are invariant under a Galilean transformation. If so, then the (Newtonian) relativity principle would hold not only for mechanics but for all of physics. That is, no inertial frame would be preferred over any other and *no type of experiment* in physics, not merely mechanical ones, carried out in a single frame would enable us to determine the velocity of our frame relative to any other frame. There would then be no preferred, or absolute, reference frame.

To see at once that the electromagnetic situation is different from the mechanical one, as far as the Galilean transformations are concerned, consider a pulse of light (i.e., an electromagnetic pulse) traveling to the right with respect to the medium through which it is propagated at a speed c. The "medium" of light propagation was given the name "ether," historically, for when the mechanical view of physics dominated physicists' thinking (late 19th century and early 20th century) it was not really accepted that an electromagnetic disturbance could be propagated in empty space. Sound waves, for example, require a medium for propagation. For simplicity, we may regard the "ether" frame, S, as an inertial one in which an observer measures the speed of light to be exactly $c = (1/\sqrt{\epsilon_0\mu_0}) = 2.997925 \times 10^8$ m/sec. In a frame S' moving at a constant speed v with respect to this ether frame, an observer would measure a different speed for the light pulse, ranging from $c + v$ to $c - v$ depending on the direction of relative motion, according to the Galilean velocity transformation.

Hence, the speed of light is certainly *not* invariant under a Galilean transformation. If these transformations really do apply to optical or electromagnetic phenomena, then there is one inertial system, and only one, in which the measured speed of light is exactly c; that is, there is a unique inertial system in which the so-called ether is at rest. We would then have a physical way of identifying an absolute (or rest) frame and of determining by optical experiments carried out in some other frame what the relative velocity of that frame is with respect to the absolute one.

A more formal way of saying this is as follows. Maxwell's equations of electromagnetism, from which we deduce the electromagnetic wave equation for example, contain the constant $c = 1/\sqrt{\mu_0\epsilon_0}$, which is identified as the velocity of propagation of a plane wave in vacuum (see *Physics*, Part II, Sec. 39-5). But such a velocity cannot be the same for observers in different inertial frames, according to the Galilean transformations, so that Maxwell's equations and therefore electromagnetic effects will probably not be the same for different inertial observers. But if we accept both the Galilean transformations and Maxwell's equations as basically correct, then it automatically follows that there exists a unique privileged frame of reference (the "ether" frame) in which Maxwell's equations are valid and in which light is propagated at a speed $c = 1/\sqrt{\mu_0 \epsilon_0}$.

The situation then seems to be as follows.* The fact that the Galilean relativity principle *does* apply to the Newtonian laws of mechanics but *not* to Maxwell's laws of electromagnetism requires us to choose the correct consequences from amongst the following possibilities.

1. A relativity principle exists for mechanics, but *not* for electrodynamics; in electrodynamics there *is* a preferred inertial frame; that is, the ether frame. Should this alternative be correct the Galilean transformations would apply and we would be able to locate the ether frame experimentally.

2. A relativity principle exists *both* for mechanics and for electrodynamics, but the laws of electrodynamics as given by Maxwell are *not* correct. If this alternative were correct, we ought to be able to perform experiments that show deviations from Maxwell's electrodynamics and reformulate the electromagnetic laws. The Galilean transformations would apply here also.

3. A relativity principle exists *both* for mechanics and for electrodynamics, but the laws of mechanics as given by Newton are *not* correct. If this alternative is the correct one, we should be able to perform experiments which show deviations from Newtonian mechanics and reformulate the mechanical laws. In that event, the correct transformation laws would not be the Galilean ones (for they are inconsistent with the invariance of Maxwell's equations) but some other ones which are consistent with classical electromagnetism and the new mechanics.

We have already indicated (Section 1-1) that Newtonian mechanics breaks down at high speeds so that the student will not be surprised to learn that alternative 3, leading to Einsteinian relativity, is the correct one. In the following sections, we shall look at the experimental bases for rejecting alternatives 1 and 2, as a fruitful prelude to finding the new relativity principle and transformation laws of alternative 3.

1.5 Attempts to Locate the Absolute Frame—The Michelson-Morley Experiment

The obvious experiment** would be one in which we can measure the speed of light in a variety of inertial systems, noting whether the measured speed is different in different systems, and if so, noting especially whether there is evidence for a single unique system—the "ether" frame—in which the speed of light is c, the value predicted from electromagnetic theory. A. A. Michelson in 1881 and Michelson and E. W. Morley in 1887 carried out such an experiment [4]. To understand the setting better, let us look a bit further into the "ether" concept.

When we say that the speed of sound in dry air at 0°C is 331.3 m/sec, we have in mind an observer, and a corresponding reference system, fixed in the air mass through which the sound wave is moving. The speed of sound

---

*The treatment here follows closely that of Ref. 2.

**Of the two famous experiments, the Trouton-Noble and the Michelson-Morley, we discuss only the latter. See Ref. 3 for a discussion of the Trouton-Noble experiment.

for observers moving with respect to this air mass is correctly given by the usual Galilean velocity transformation Eq. 1-1. However, when we say that the speed of light in a vacuum is 2.997925×10^8 m/sec $(= 1/\sqrt{\mu_0 \epsilon_0})$ it is not at all clear what reference system is implied. A reference system fixed in the medium of propagation of light presents difficulties because, in contrast to sound, no medium seems to exist. However, it seemed inconceivable to 19th century physicists that light and other electromagnetic waves, in contrast to all other kinds of waves, could be propagated without a medium. It seemed to be a logical step to postulate such a medium, called the ether, even though it was necessary to assume unusual properties for it, such as zero density and perfect transparency, to account for its undetectability. The ether was assumed to fill all space and to be the medium with respect to which the speed c applies. It followed then that an observer moving through the ether with velocity **v** would measure a velocity **c'** for a light beam, where **c'** = **c** − **v**. It was this result that the Michelson-Morley experiment was designed to test.

If an ether exists, the spinning and rotating earth should be moving through it. An observer on earth would sense an "ether wind," whose velocity is **v** relative to the earth. If we were to assume that v is equal to the earth's orbital speed about the sun, about 30 km/sec, then $v/c \approx 10^{-4}$. Optical experiments, which were accurate to the first order in v/c, were not able to detect the absolute motion of the earth through the ether, but Fresnel (and later Lorentz) showed how this result could be interpreted in terms of an ether theory. This interpretation had difficulties, however, so that the issue was not really resolved satisfactorily with first-order experiments (experiments whose results depend only on the first power of the ratio v/c). It was generally agreed that an unambiguous test of the ether hypothesis would require an experiment that measured the "second-order" effect, that is, one that measured $(v/c)^2$. The first-order effect is not large to begin with ($v/c = 10^{-4}$, an effect of one part in 10,000) but the second-order effect is really very small ($v^2/c^2 = 10^{-8}$, an effect of one part in 100 million).

It was A. A. Michelson (1852–1931) who invented the optical interferometer whose remarkable sensitivity made such an experiment possible. Michelson first performed the experiment in 1881, and then—in 1887, in collaboration with E. W. Morley—carried out the more precise version of the investigation that was destined to lay the experimental foundations of relativity theory. For his invention of the interferometer and his many optical experiments, Michelson was awarded the Nobel Prize in Physics in 1907, the first American to be so honored.

Let us now describe the Michelson-Morley experiment. The Michelson interferometer (Fig. 1-4) is fixed on the earth. If we imagine the "ether" to be fixed with respect to the sun, then the earth (and interferometer) moves through the ether at a speed of 30 km/sec, in different directions in different seasons (Fig. 1-5). For the moment, neglect the earth's spinning motion. The beam of light (plane waves, or parallel rays) from the laboratory source S (fixed with respect to the instrument) is split by the partially silvered mirror M into two coherent beams, beam 1 being transmitted through M and beam 2 being reflected off of M. Beam 1 is reflected back to M by mirror M_1 and beam 2 by mirror M_2. Then the returning beam 1 is partially reflected and

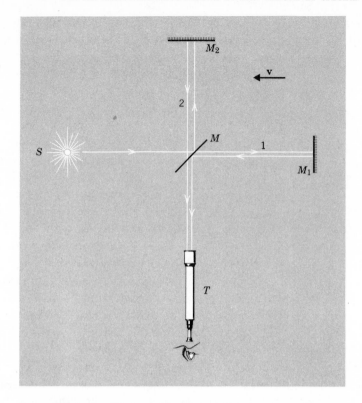

Figure 1-4. A simplified version of the Michelson interferometer showing how the beam from the source S is split into two beams by the partially silvered mirror M. The beams are reflected by mirrors 1 and 2, returning to the partially silvered mirror. The beams are then transmitted to the telescope T where they interfere, giving rise to a fringe pattern. In this figure, **v** *is the velocity of the ether with respect to the interferometer.*

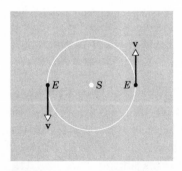

Figure 1-5. The earth E moves at an orbital speed of 30 km/sec along its nearly circular orbit about the sun S, reversing the direction of its velocity every six months.

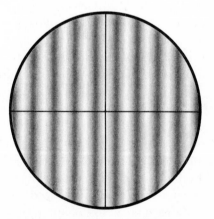

Figure 1-6. A typical fringe system seen through the telescope T when M_1 and M_2 are not quite at right angles.

the returning beam 2 is partially transmitted by M back to a telescope at T where they interfere. The interference is constructive or destructive depending on the phase difference of the beams. The partially silvered mirror surface M is inclined at $45°$ to the beam directions. If M_1 and M_2 are very nearly (but not quite) at right angles, we shall observe a fringe system in the telescope (Fig. 1-6) consisting of nearly parallel lines, much as we get from a thin wedge of air between two glass plates.

Let us compute the phase difference between the beams 1 and 2. This difference can arise from two causes, the different path lengths traveled, l_1 and l_2, and the different speeds of travel with respect to the instrument because of the "ether wind" v. The second cause, for the moment, is the crucial one. The different speeds are much like the different cross-stream and up-and-down-stream speeds with respect to shore of a swimmer in a moving stream. The time for beam 1 to travel from M to M_1 and back is

$$t_1 = \frac{l_1}{c - v} + \frac{l_1}{c + v} = l_1\left(\frac{2c}{c^2 - v^2}\right) = \frac{2l_1}{c}\left(\frac{1}{1 - v^2/c^2}\right)$$

for the light, whose speed is c in the ether, has an "upstream" speed of $c - v$ with respect to the apparatus and a "downstream" speed of $c + v$. The path of beam 2, traveling from M to M_2 and back, is a cross-stream path through the ether, as shown in Fig. 1-7, enabling the beam to return to the (advancing)

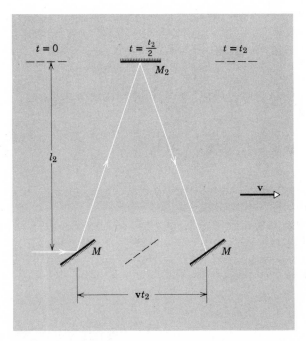

Figure 1-7. The cross-stream path of beam 2. The mirrors move through the "ether" at a speed v, the light moving through the "ether" at speed c. Reflection from the moving mirror automatically gives the cross-stream path. In this figure, **v** *is the velocity of the interferometer with respect to the "ether."*

mirror M. The transit time is given by

$$2\left[l_2^2 + \left(\frac{vt_2}{2}\right)^2\right]^{1/2} = ct_2$$

or

$$t_2 = \frac{2l_2}{\sqrt{c^2 - v^2}} = \frac{2l_2}{c}\frac{1}{\sqrt{1 - v^2/c^2}}.$$

The calculation of t_2 is made in the ether frame, that of t_1 in the frame of the apparatus. Because time is an absolute in classical physics, this is perfectly acceptable classically. Note that both effects are second-order ones $(v^2/c^2 \approx 10^{-8})$ and are in the same direction (they *increase* the transit time over the case $v = 0$). The difference in transit times is

$$\Delta t = t_2 - t_1 = \frac{2}{c}\left[\frac{l_2}{\sqrt{1 - v^2/c^2}} - \frac{l_1}{1 - v^2/c^2}\right]$$

Suppose that the instrument is rotated through 90°, thereby making l_1 the cross-stream length and l_2 the downstream length. If the corresponding times are now designated by primes, the same analysis as above gives the transit-time difference as

$$\Delta t' = t_2' - t_1' = \frac{2}{c}\left[\frac{l_2}{1 - v^2/c^2} - \frac{l_1}{\sqrt{1 - v^2/c^2}}\right].$$

Hence, *the rotation changes the differences* by

$$\Delta t' - \Delta t = \frac{2}{c}\left[\frac{l_2 + l_1}{1 - v^2/c^2} - \frac{l_2 + l_1}{\sqrt{1 - v^2/c^2}}\right].$$

Using the binomial expansion* and dropping terms higher than the second-order, we find

$$\Delta t' - \Delta t \simeq \frac{2}{c}(l_1 + l_2)\left[1 + \frac{v^2}{c^2} - 1 - \frac{1}{2}\frac{v^2}{c^2}\right] = \left(\frac{l_1 + l_2}{c}\right)\frac{v^2}{c^2}.$$

Therefore, the rotation should cause a shift in the fringe pattern, since it changes the phase relationship between beams 1 and 2.

If the optical path difference between the beams changes by one wavelength, for example, there will be a shift of one fringe across the crosshairs of the viewing telescope. Let ΔN represent the number of fringes moving past the crosshairs as the pattern shifts. Then, if light of wavelength λ is used,

*In these cases the binomial expansion gives

$$\frac{1}{(1 - v^2/c^2)} = 1 + \frac{v^2}{c^2} + \left(\frac{v^2}{c^2}\right)^2 + \left(\frac{v^2}{c^2}\right)^3 + \cdots.$$

and

$$\frac{1}{(1 - v^2/c^2)^{1/2}} = 1 + \frac{1}{2}\left(\frac{v^2}{c^2}\right) + \frac{3}{8}\left(\frac{v^2}{c^2}\right)^2 + \frac{5}{16}\left(\frac{v^2}{c^2}\right)^3 + \cdots.$$

See also Problem 9.

so that the period of one vibration is $T = 1/\nu = \lambda/c$,

$$\Delta N = \frac{\Delta t' - \Delta t}{T} \cong \frac{l_1 + l_2}{cT}\frac{v^2}{c^2} = \frac{l_1 + l_2}{\lambda}\frac{v^2}{c^2}. \qquad (1\text{-}8)$$

Michelson and Morley were able to obtain an optical path length, $l_1 + l_2$, of about 22 m. In their experiment the arms were of (nearly) equal length, that is, $l_1 = l_2 = l$, so that $\Delta N = (2l/\lambda)(v^2/c^2)$. If we choose $\lambda = 5.5 \times 10^{-7}$ m and $v/c = 10^{-4}$, we obtain, from Eq. 1-8,

$$\Delta N = \frac{22\ \text{m}}{5.5 \times 10^{-7}\ \text{m}} 10^{-8} = 0.4,$$

or a shift of four-tenths a fringe!

Michelson and Morley mounted the interferometer on a massive stone slab for stability and floated the apparatus in mercury so that it could be rotated smoothly about a central pin. In order to make the light path as long as possible, mirrors were arranged on the slab to reflect the beams back and forth through eight round trips. The fringes were observed under a continuous rotation of the apparatus and a shift as small as $\frac{1}{100}$ of a fringe definitely could have been detected (see Fig. 1-8). Observations were made day and night (as the earth spins about its axis) and during all seasons of the year

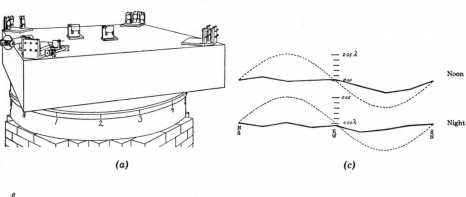

(a)   (c)

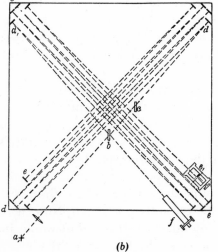

(b)

Figure 1-8. (*a*) Mounting of the Michelson-Morley apparatus. (*b*) Plan view. (*c*) Observed results. The broken solid lines show the observed fringe shift in the Michelson-Morley experiment as a function of the angle of rotation of the interferometer. The smooth dashed curves—which should be *multiplied by a factor of 8* to bring it to the proper scale—show the fringe shift predicted by the ether hypothesis. (From "On the Relative Motion of the Earth and the Luminiferous Aether" by Albert A. Michelson and Edward W. Morley, *The London, Edinburgh, and Dublin Philosophical Magazine and Journal of Science,* December 1887.)

Table 1-1   TRIALS OF THE MICHELSON-MORLEY EXPERIMENT

Observer	Year	Place	*l* Meters	Fringe Shift Predicted by Ether Theory	Upper Limit of Observed Fringe Shift
Michelson	1881	Potsdam	1.2	0.04	0.02
Michelson and Morley	1887	Cleveland	11.0	0.40	0.01
Morley and Miller	1902–1904	Cleveland	32.2	1.13	0.015
Miller	1921	Mt. Wilson	32.0	1.12	0.08
Miller	1923–1924	Cleveland	32.0	1.12	0.030
Miller (sunlight)	1924	Cleveland	32.0	1.12	0.014
Tomaschek (starlight)	1924	Heidelberg	8.6	0.3	0.02
Miller	1925–1926	Mt. Wilson	32.0	1.12	0.088
Kennedy	1926	Pasadena and Mt. Wilson	2.0	0.07	0.002
Illingworth	1927	Pasadena	2.0	0.07	0.0004
Piccard and Stahel	1927	Mt. Rigi	2.8	0.13	0.006
Michelson et al	1929	Mt. Wilson	25.9	0.9	0.010
Joos	1930	Jena	21.0	0.75	0.002

Source. From Shankland, McCuskey, Leone, and Kuerti, *Rev. Mod. Phys.,* **27,** 167 (1955).

(as the earth rotates about the sun), but the expected fringe shift was not observed. Indeed, the experimental conclusion was that *there was no fringe shift at all.*

This null result ($\Delta N = 0$) was such a blow to the ether hypothesis that the experiment was repeated by many workers over a 50-year period. The null result was amply confirmed (see Table 1-1) and provided a great stimulus to theoretical and experimental investigation. In 1958 J. P. Cedarholm, C. H. Townes et al [5] carried out an "ether-wind" experiment using microwaves in which they showed that if there is an ether and the earth is moving through it, the earth's speed with respect to the ether would have to be less than $\frac{1}{1000}$ of the earth's orbital speed. This is an improvement of 50 in precision over the best experiment of the Michelson-Morley type. The null result is well established.

The student should note that the Michelson-Morley experiment depends essentially on the 90° rotation of the interferometer, that is, on interchanging the roles of l_1 and l_2, as the apparatus moves with a speed v through an "ether." In predicting an expected fringe shift, we took **v** to be the earth's velocity with respect to an ether fixed with the sun. However, the solar system itself might be in motion with respect to the hypothetical ether. Actually, the experimental results themselves determine the earth's speed with respect to an ether, if indeed there is one, and these results give $v = 0$. Now, if at some time the velocity were zero in such an ether, no fringe shift would be expected, of course. But the velocity cannot always be zero, since the velocity of the apparatus is *changing* from day to night (as the earth spins) and from season to season (as the earth rotates about the sun). Therefore, the experiment does not depend solely on an "absolute" velocity of the earth through an ether, but also depends on the changing velocity of the earth with respect to the "ether." Such a changing motion through the "ether" would be easily detected and measured by the precision experiments, if there were an ether frame. The null result seems to rule out an ether (absolute) frame.

One way to interpret the null result of the Michelson-Morley experiment is to conclude simply that the measured speed of light is the same, that is, c, for all directions in every inertial system. For this fact would lead to $\Delta N = 0$ in the (equal arm) experiment, the "downstream" and "cross-stream" speeds being c, rather than $|c + v|$, in any frame. However, such a conclusion, being incompatible with the Galilean (velocity) transformations, seemed to be too drastic philosophically at the time. If the measured speed of light did not depend on the motion of the observer, all inertial systems would be equivalent for a propagation of light and there could be no experimental evidence to indicate the existence of a unique inertial system, that is, the ether. Therefore, to "save the ether" and still explain the Michelson-Morley result, scientists suggested alternative hypotheses. We explore these alternatives in succeeding sections.

1.6 Attempts to Preserve the Concept of a Preferred Ether Frame—The Lorentz-Fitzgerald Contraction Hypothesis

Fitzgerald (in 1892) proposed a hypothesis, which was elaborated upon by Lorentz, to explain the Michelson-Morley null result and still to retain the concept of a preferred ether frame. Their hypothesis was that all bodies are contracted in the direction of motion relative to the stationary ether by a factor $\sqrt{1 - v^2/c^2}$. For convenience, let the ratio v/c be represented by the symbol β, so that this factor may be written as $\sqrt{1 - \beta^2}$. Now, if $l°$ represents the length of a body at rest with respect to the ether (its rest length) and l its length when in motion with respect to the ether, then in the Michelson-Morley experiment

$$l_1 = l_1° \sqrt{1 - \beta^2} \qquad \text{and} \qquad l_2 = l_2°.$$

This last result follows from the fact that in the hypothesis it was assumed that lengths at right angles to the motion are unaffected by the motion. Then

$$\Delta t = \frac{2}{c} \frac{1}{\sqrt{1 - \beta^2}} (l_2° - l_1°)$$

and, on 90° rotation, (1-9)

$$\Delta t' = \frac{2}{c} \frac{1}{\sqrt{1 - \beta^2}} (l_2° - l_1°)$$

Hence, no fringe shift should be expected on *rotation* of the interferometer, for $\Delta t' - \Delta t = 0$.

Lorentz was able to account for such a contraction in terms of his electron theory of matter, but the theory was elaborate and somewhat contrived and other results predicted from it could not be found experimentally. As for the interferometer result of the contraction hypothesis, it too can be demolished as a correct explanation. Recall that, in the original experiment, the arms were of (nearly) equal length ($l_1 = l_2 = l$). Consider now an interferometer in which $l_1 \neq l_2$. In that case, *even including the Lorentz contraction effect,* we should expect a fringe shift when the velocity of the interferometer changes with respect to the ether from v to v'. The predicted shift in fringes (to second-order terms; see Problem 11) is

$$\Delta N = \frac{l_1^{\circ} - l_2^{\circ}}{\lambda} \left(\frac{v^2}{c^2} - \frac{v'^2}{c^2} \right). \tag{1-10}$$

Kennedy and Thorndike [6], using an interferometer with unequal arms (the path difference was about 16 cm, as great as permitted by coherence of the source), carried out the appropriate experiment. Although the difference $(v^2 - v'^2)/c^2$ should change as a result of the earth's spin (the biggest change occurring in twelve hours) and the earth's rotation (the biggest change occurring in six months), neither effect was observed (i.e., $\Delta N = 0$) in direct contradiction to the contraction hypothesis.

1.7 Attempts to Preserve the Concept of a Preferred Ether Frame—The Ether-Drag Hypothesis

Another idea advanced to retain the notion of an ether was that of "ether drag." This hypothesis assumed that the ether frame was attached to all bodies of finite mass, that is, dragged along with such bodies. The assumption of such a "local" ether would automatically give a null result in the Michelson-Morley experiment. Its attraction lay in the fact that it did not require modification of classical mechanics or electromagnetism. However, there were two well-established effects which contradicted the ether-drag hypothesis: stellar aberration and the Fizeau convection coefficient. Let us consider these effects now, since we must explain them eventually by whatever theory we finally accept.

The aberration of light was first reported by Bradley [see Ref. 7] in 1727. He observed that (with respect to astronomical coordinates fixed with respect to the earth) the stars appear to move in circles, the angular diameter of these circular orbits being about 41 seconds of arc. This can be understood as follows. Imagine that a star is directly overhead so that a telescope would have to be pointed straight up to see it if the earth were at rest in the ether. That is (see Fig. 1-9a), the rays of light coming from the star would proceed straight down the telescope tube. Now, imagine that the earth is moving to the right through the ether with a speed v. In order for the rays to pass down the telescope tube without hitting its sides—that is, in order to see the star—we would have to tilt the telescope as shown in Fig. 1-9b. The light proceeds straight down in the ether (as before) but, during the time Δt that the light travels the vertical distance $l = c \, \Delta t$ from the objective lens to the eyepiece, the telescope has moved a distance $v \, \Delta t$ to the right. The eyepiece, at the time the ray leaves the telescope, is on the same vertical line as the objective lens was at the time the ray entered the telescope. From the point

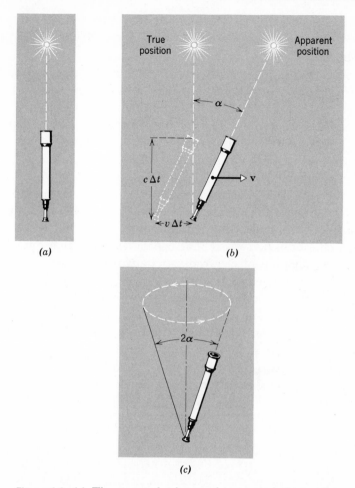

Figure 1-9. (*a*) The star and telescope have no relative motion (i.e., both are at rest in the ether); the star is directly overhead. (*b*) The telescope now moves to the right at speed *v* through the ether, it must be tilted at an angle α (greatly exaggerated in the drawing) from the vertical to see the star, whose apparent position now differs from its true position. ("True" means with respect to the sun, i.e., with respect to an earth that has no motion relative to the sun.) (*c*) A cone of abberation of angular diameter 2α is swept out by the telescope axis during the year.

of view of the telescope, the ray travels along the axis from objective lens to eyepiece. The angle of tilt of the telescope, α, is given by

$$\tan \alpha = \frac{v\,\Delta t}{c\,\Delta t} = \frac{v}{c}. \tag{1-11}$$

It was known that the earth goes around the sun at a speed of about 30 km/sec, so that with $c = 3 \times 10^5$ km/sec, we obtain an angle α = 20.5 sec of arc. The earth's motion is nearly circular so that the direction of aberration reverses every six months, the telescope axis tracing out a cone of aberration during the year (Fig. 1-9c). The angular diameter of the cone, or of the observed circular path of the star, would then be 2α = 41 sec of arc, in excellent agreement with the observations. For stars not directly overhead, the analysis, although more involved, is similar and identical in principle (see Problem 12).

The important thing we conclude from this agreement is that the ether is *not* dragged around with the earth. If it were, the ether would be at rest with respect to the earth, the telescope would not have to be tilted, and there would be no aberration at all. That is, the ether would be moving (with the earth) to the right with speed v in Fig. 1-9c, so there would be no need to correct for the earth's motion through the ether; the light ray would be swept along with the ether just as a wind carries a sound wave with it. Hence, *if* there is an ether, it is *not* dragged along by the earth but, instead, the earth moves freely through it. Therefore, we cannot explain the Michelson-Morley result by means of an ether-drag hypothesis.

Another well-established effect that contradicts the ether-drag hypothesis involves the propagation of electromagnetic waves in moving media. J. A. Fresnel, in 1817, predicted that light would be partially dragged along by a moving medium and derived an exact formula for the effect on the basis of an ether hypothesis. The effect was confirmed experimentally by Fizeau in 1851. The set-up of the Fizeau experiment is shown diagrammatically in Fig. 1-10. Light from the source S falls on a partially silvered mirror M which splits the beam into two parts. One part is transmitted to mirror M_1 and proceeds in a counterclockwise sense back to M, after reflections at M_1, M_2, and M_3. The other part is reflected to M_3 and proceeds in a clockwise sense

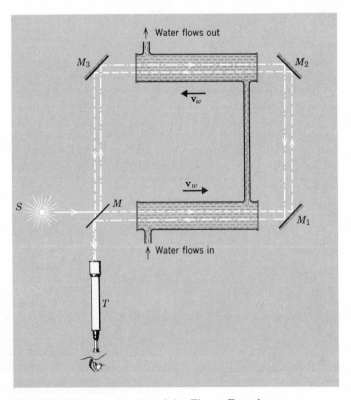

Figure 1-10. Schematic view of the Fizeau Experiment.

back to M, after reflections at M_3, M_2, M_1. At M, part of the returning first beam is transmitted and part of the returning second beam is reflected to the telescope T. Interference fringes, representing optical path differences of the beams, will be seen in the telescope. Water flows through the tubes (which have flat glass end sections) as shown, so that one light beam always travels in the direction of flow and the other always travels opposite to the direction of flow. The flow of water can be reversed, of course, but outside the tubes conditions remain the same for each beam.

Let the apparatus be our S-frame. In this laboratory frame the velocity of light in still water is c/n and the velocity of the water is v_w. Does the flow of water, the medium through which the light passes, affect the velocity of light measured in the laboratory? According to Fresnel the answer is yes. The velocity of light, v, in a body of refractive index, n, moving with a velocity v_w relative to the observer (i.e., to the frame of reference S in which the free-space velocity of light would be c) is given by Fresnel as

$$v = \frac{c}{n} \pm v_w \left(1 - \frac{1}{n^2} \right). \tag{1-12}$$

The factor $(1 - 1/n^2)$ is called the Fresnel drag coefficient. The speed of light is changed from the value c/n because of the motion of the medium but, because the factor is less than unity, the change (increase or decrease) of speed is less than the speed v_w of the medium—hence the term "drag." For yellow sodium light in water, for example, the speed increase (or decrease) is 0.565 v_w. Notice that for $n = 1$ ("a moving vacuum") Eq. 1-12 reduces plausibly to $v = c$.

This result can be understood by regarding the light as being carried along both by the refractive medium and by the ether that permeates it. Then, with the ether at rest and the refractive medium moving through the ether, the light will act to the rest observer as though only a part of the velocity of the medium were added to it. The result can be derived directly from electromagnetic theory.

In Fizeau's experiment, the water flowed through the tubes at a speed of about 7 m/sec. Fringe shifts were observed from the zero flow speed to flow speeds of 7 m/sec, and on reversing the direction of flow. Fizeau's measurements confirmed the Fresnel prediction. The experiment was repeated by Michelson and Morley in 1886 and by P. Zeeman and others after 1914 under conditions allowing much greater precision, again confirming the Fresnel drag coefficient.

◗ **Example 5.** In Fizeau's experiment, the approximate values of the parameters were as follows: $l = 1.5$ m, $n = 1.33$, $\lambda = 5.3 \times 10^{-7}$ m, and $v_w = 7$ m/sec. A shift of 0.23 fringe was observed from the case $v_w = 0$. Calculate the drag coefficient and compare it with the predicted value.

Let d represent the drag coefficient. The time for beam 1 to traverse the water is then

$$t_1 = \frac{2l}{(c/n) - v_w d}$$

and for beam 2

$$t_2 = \frac{2l}{(c/n) + v_w d}.$$

Hence,

$$\Delta t = t_1 - t_2 = \frac{4 l v_w d}{(c/n)^2 - v_w^2 d^2} \cong \frac{4 l n^2 v_w d}{c^2}$$

The period of vibration of the light is $T = \lambda/c$ so that

$$\Delta N \cong \frac{\Delta t}{T} = \frac{4 l n^2 v_w d}{\lambda c}$$

and, with the values above, we obtain

$$d = \frac{\lambda c \, \Delta N}{4 l n^2 v_w} = 0.47.$$

The Fresnel prediction (see Eq. 1-12) is

$$d = 1 - \frac{1}{n^2} = 0.44. \quad \blacktriangleleft$$

If the ether were dragged along with the water, the velocity of light in the laboratory frame, using the Galilean ideas, would have been $(c/n) + v_w$ in one tube and $(c/n) - v_w$ in the other tube. Instead, the Fizeau experiment, as we have seen, is interpreted most simply in terms of no ether drag at all, either by the apparatus or the water moving through it, and a partial drag due to the motion of the refractive medium. Indeed, the aberration experiment, when done with a telescope filled with water (see Question 15), leads to exactly the same result and interpretation. Hence, the ether-drag hypothesis is contradicted by the facts.

There appears to be no acceptable experimental basis then for the idea of an ether, that is, for a preferred frame of reference. This is true whether we choose to regard the ether as stationary or as dragged along. We must now face the alternative that a principle of relativity is valid in electrodynamics as well as in mechanics. If this is so, then either electrodynamics must be modified, so that it is consistent with the classical relativity principle, or else we need a new relativity principle that is consistent with electrodynamics, in which case classical mechanics will need to be modified.

1.8 Attempts to Modify Electrodynamics

Let us consider now the attempts to modify the laws of electromagnetism. A possible interpretation of the Michelson-Morley result is that the velocity of light has the same value in all inertial frames. If this is so, then the velocity of light surely cannot depend on the velocity of the light source relative to the observer. However, this principle of the invariance of the velocity of light contradicts the classical relativity principle. If we wish to avoid such an interpretation of the Michelson-Morley experiment, one modification of electromagnetism that suggests itself is to assume that the velocity of a light wave *is* connected with the motion of the source rather than with an ether.

The various theories that are based on this assumption are called *emission* theories. Common to them all is the hypothesis that the velocity of light is c relative to the original source and that this velocity is independent of the state of motion of the medium transmitting the light. This would automatically explain the null result of the Michelson-Morley experiment. The theories differ in their predictions as to what the velocity of light becomes on reflection from a moving mirror.* Nevertheless, all emission theories are contradicted directly by two types of experiment. The first is typified by the de Sitter observations on double (or binary) stars (see Ref. 9 and Problem 14), the second by a Michelson-Morley experiment using an extraterrestrial light source.

Two stars that are close to one another and move in orbits about their common center of mass are called double stars. Imagine the orbits to be circles. Now assume that the velocity of the light by which we see them through the empty space is equal to $c + v_s$, where v_s is the component of the velocity of the source relative to the observer, at the time the light is emitted, along the line from the source to the observer. Then, the time for light to reach the earth from the approaching star would be smaller than that from the receding star. As a consequence, the circular orbits of double stars should appear to be eccentric as seen from earth. But measurements show no such eccentricities in the orbits of double stars observed from earth.

The results are consistent with the assumption that the velocity of the light is independent of the velocity of the source.** De Sitter's conclusion was that, if the velocity of light is not equal to c but is equal instead to $c + kv_s$, then k experimentally must be less than 2×10^{-3}. More recent experiments [11], using fast-moving terrestrial sources, confirm the conclusion that the velocity of electromagnetic radiation is independent of the velocity of the source. In such a recent experiment (1964), measurements were made of the speed of electromagnetic radiation emitted from the decay of rapidly moving π° mesons produced in the CERN synchrotron. The mesons had energies greater than 6 GeV ($v_s = 0.99975c$) and the speed of the γ-radiation emitted from these fast-moving sources was measured absolutely by timing over a known distance. The result corresponded to a value of k equal to $(-3 \pm 13) \times 10^{-5}$.

The Michelson-Morley experiment, using an extraterrestrial source, has been performed by R. Tomaschek, who used starlight, and D. C. Miller, who used sunlight [12]. If the source velocity (due to rotational and translational motions relative to the interferometer) affects the velocity of light, we should observe complicated fringe-pattern changes. No such effects were observed in either of the experiments.

We saw earlier that an ether hypothesis is untenable. Now we are forced by experiment to conclude further that the laws of electrodynamics are correct and do not need modification. The speed of light (i.e., electromagnetic radiation) is the same in all inertial systems, independent of the relative motion of source and observer. Hence, a relativity principle, applicable both to

---

*The *original source* theory assumes that the velocity remains c relative to the source; the *ballistic* theory assumes that the velocity becomes c relative to the mirror; and *the new source* theory assumes that the velocity becomes c relative to the mirror image of the source. See Refs. 8.

**For an analysis and a discussion of alternate interpretations of the de Sitter experiment, see Ref. 10.

mechanics *and* to electromagnetism, is operating. Clearly, it cannot be the Galilean principle, since that required the speed of light to depend on the relative motion of source and observer. We conclude that the Galilean transformations must be replaced and, therefore, the basic laws of mechanics, which were consistent with those transformations, need to be modified so that they will be consistent with the new transformations.

1.9   The Postulates of Special Relativity Theory

In 1905, before many of the experiments we have discussed were actually performed (see Question 18), Albert Einstein (1879–1955), apparently unaware of several earlier important papers on the subject, provided a solution to the dilemma facing physics. In his paper "On the Electrodynamics of Moving Bodies" [13], Einstein wrote ". . . no properties of observed facts correspond to a concept of absolute rest; . . . for all coordinate systems for which the mechanical equations hold, the equivalent electrodynamical and optical equations hold also. . . . In the following we make these assumptions (which we shall subsequently call the Principle of Relativity) and introduce the further assumption—an assumption which is at the first sight quite irreconcilable with the former one—that light is propagated in vacant space, with a velocity c which is independent of the nature of motion of the emitting body. These two assumptions are quite sufficient to give us a simple and consistent theory of electrodynamics of moving bodies on the basis of the Maxwellian theory for bodies at rest."

We can rephrase these assumptions of Einstein as follows.

1. *The laws of physics are the same in all inertial systems. No preferred inertial system exists.* (The Principle of Relativity.)

2. *The speed of light in free space has the same value c in all inertial systems.* (The Principle of the Constancy of the Speed of Light.)

Einstein's relativity principle goes beyond the Newtonian relativity principle, which dealt only with the laws of mechanics, to include *all* the laws of physics. It states that it is impossible by means of *any* physical measurements to designate an inertial system as intrinsically stationary or moving; we can only speak of the *relative* motion of two systems. Hence, no physical experiment of *any* kind made entirely *within* an inertial system can tell the observer what the motion of his system is with respect to any other inertial system. The second principle above, which flatly contradicts the Galilean velocity transformation (Eq. 1-5), is clearly consistent with the Michelson-Morley (and subsequent) experiments. The entire special theory of relativity is derived directly from these two assumptions. Their simplicity, boldness, and generality are characteristic of Einstein's genius. The success of his theory, as indeed of any theory, can only be judged by comparison with experiment. It not only was able to explain all the existing experimental results but predicted new effects which were confirmed by later experiments. No experimental objection to Einstein's special theory of relativity has yet been found.

In Table 1-2 we list the seven theories proposed at various times and compare their predictions of the results of thirteen crucial experiments, old and new. Notice that only the special theory of relativity is in agreement

Table 1-2 EXPERIMENTAL BASIS FOR THE THEORY OF SPECIAL RELATIVITY

Theory	Light Propagation Experiments							Experiments from Other Fields					
	Aberration	Fizeau convection coefficient	Michelson-Morley	Kennedy-Thorndike	Moving sources and mirrors	De Sitter spectroscopic binaries	Michelson-Morley, using sunlight	Variation of mass with velocity	General mass-energy equivalence	Radiation from moving charges	Meson decay at high velocity	Trouton-Noble	Unipolar induction, using permanent magnet
Ether theories — Stationary ether, no contraction	A	A	D	D	A	A	D	D	N	A	N	D	D
Stationary ether, Lorentz contraction	A	A	A	D	A	A	A	A	N	A	N	A	D
Ether attached to ponderable bodies	D	D	A	A	A	A	A	D	N	N	N	A	N
Emission theories — Original source	A	A	A	A	A	D	D	N	N	D	N	N	N
Ballistic	A	N	A	A	D	D	D	N	N	D	N	N	N
New source	A	N	A	A	D	D	A	N	N	D	N	N	N
Special theory of relativity	A	A	A	A	A	A	A	A	A	A	A	A	A

Legend. A, the theory agrees with experimental results.
　　　D, the theory disagrees with experimental results.
　　　N, the theory is not applicable to the experiment.
Source. From Panofsky and Phillips, *Classical Electricity and Magnetism* (2nd ed.), Addison-Wesley, New York (1962).

with *all* the experiments listed. We have already commented on the successes and failures of the ether and emission theories with most of the light-propagation experiments and it remains for us to show how special relativity accounts for their results. In addition, several experiments from other fields—some suggested by the predictions of relativity and in flat contradiction to Newtonian mechanics—remain to be examined. What emerges from this comparative preview is the compelling *experimental* basis of special relativity theory. It alone is in accord with the real world of experimental physics.

As is often true in the aftermath of a great new theory, it seemed obvious to many in retrospect that the old ideas had to be wrong. For example, Herman Bondi [15] has said:

"The special theory of relativity is a necessary consequence of any assertion that the unity of physics is essential, for it would be intolerable for all inertial systems to be equivalent from a dynamical point of view yet distinguishable by optical measurements. It now seems almost incredible that the possibility of such a discrimination was taken for granted in the nineteenth century, but at the same time it was not easy to see what was more important—the universal validity of the Newtonian principle of relativity or the absolute nature of time."

It was his preoccupation with the nature of time that led Einstein to his revolutionary proposals. We shall see later how important a clear picture of the concept of time was to the development of relativity theory. However, the program of the theory, in terms of our discussions in this chapter, should now be clear. First, we must obtain equations of transformation between two uniformly moving (inertial) systems which will keep the velocity of light constant. Second, we must examine the laws of physics to check whether or not they keep the same form (i.e., are invariant) under this transformation. Those laws that are not invariant will need to be generalized so as to obey the principle of relativity.

The new equations of transformation obtained in this way by Einstein are known for historical reasons as a Lorentz transformation. We have seen (Section 1-3) that Newton's equation of motion is invariant under a Galilean transformation, which we now know to be incorrect. It is likely then that Newton's laws—and perhaps other commonly accepted laws of physics—will not be invariant under a Lorentz transformation. In that case, they must be generalized. We expect the generalization to be such that the new laws will reduce to the old ones for velocities much less than that of light, for in that range both the Galilean transformation and Newton's laws are at least approximately correct.

In Table 1-3, for perspective, we compare relativity theory to the older emission and ether theories in terms of their basic assumptions and conclusions.

1.10   Einstein and the Origin of Relativity Theory

It is so fascinating a subject that one is hard-pressed to cut short a discussion of Albert Einstein, the person. Common misconceptions of the man, who quite properly symbolized for his generation the very height of intellect, might be shattered by such truths as these: Einstein's parents feared for a while that he might be mentally retarded for he learned to speak much later than customary; one of his teachers said to him "You will never amount to anything, Einstein," in despair at his daydreaming and his negative attitude toward formal instruction; he failed to get a high-school diploma and, with no job prospects, at the age of fifteen he loafed like a "model dropout"; Einstein's first attempt to gain admission to a polytechnic institute ended when he failed to pass an entrance examination; after gaining admittance he cut most of the lectures and, borrowing a friend's class notes, he crammed intensively for two months before the final examinations.

Table 1-3 BASIC ASSUMPTIONS AND CONCLUSIONS OF ALTERNATIVE THEORIES

	Emission Theory	Classical Ether Theory	Special Theory of Relativity
Reference system	No special reference system	Stationary ether is special reference system	No special reference system
Velocity dependence	The velocity of light depends on the motion of the source	The velocity of light is independent of the motion of the source	The velocity of light is independent of the motion
Space-time connection	Space and time are independent	Space and time are independent	Space and time are interdependent
Transformation equations	Inertial frames in relative motion are connected by a Galilean transformation	Inertial frames in relative motion are connected by a Galilean transformation	Inertial frames in relative motion are connected by a Lorentz transformation

Source. From Panofsky and Phillips, *Classical Electricity and Magnetism* (2nd Ed.), Addison-Wesley, New York (1962).

He later said of this ". . . after I had passed the final examination, I found the consideration of any scientific problem distasteful to me for an entire year." It was not until two years after his graduation that he got a steady job, as a patent examiner in the Swiss Patent Office at Berne; Einstein was very interested in technical apparatus and instruments, but—finding he could complete a day's work in three or four hours—he secretly worked there, as well as in his free time, on the problems in physics which puzzled him. And so it goes.*

The facts above are surprising only when considered in isolation, of course. Einstein simply could not accept the conformity required of him, whether in educational, religious, military, or governmental institutions. He was an avid reader who pursued his own intellectual interests, had a great curiosity about nature, and was a genuine "freethinker" and independent spirit. As Martin Klein (Ref. 16) points out, what is really surprising about Einstein's early life is that none of his "elders" recognized his genius.

But such matters aside, let us look now at Einstein's early work. It is appropriate to quote here from Martin Klein [16].

"In his spare time during those seven years at Berne, the young patent examiner wrought a series of scientific miracles; no weaker word is adequate. He did nothing less than to lay out the main lines along which twentieth-century theoretical physics has developed. A very brief list will have to suffice. He began by working out the subject of statistical mechanics quite independently and without knowing of the work of J. Willard Gibbs. He also took this subject seriously in a way that neither Gibbs nor Boltzman had ever done, since he used it to

*See Refs. 16–21 for some rewarding articles and books about Einstein.

give the theoretical basis for a final proof of the atomic nature of matter. His reflections on the problems of the Maxwell-Lorentz electrodynamics led him to create the special theory of relativity. Before he left Berne he had formulated the principle of equivalence and was struggling with the problems of gravitation which he later solved with the general theory of relativity. And, as if these were not enough, Einstein introduced another new idea into physics, one that even he described as 'very revolutionary,' the idea that light consists of particles of energy. Following a line of reasoning related to but quite distinct from Planck's, Einstein not only introduced the light quantum hypothesis, but proceeded almost at once to explore its implications for phenomena as diverse as photochemistry and the temperature dependence of the specific heat of solids.

"What is more, Einstein did all this completely on his own, with no academic connections whatsoever, and with essentially no contact with the elders of his profession."

The discussion thus far emphasizes Einstein's independence of other contemporary workers in physics. Also characteristic of his work is the fact that he always made specific predictions of possible experiments to verify his theories. In 1905, at intervals of less than eight weeks, Einstein sent to the *Annalen der Physik* three history-making papers. The first paper [22] on the quantum theory of light included an explanation of the photoelectric effect. The suggested experiments, which gave the proof of the validity of Einstein's equations, were successfully carried out by Robert A. Millikan nine years later! The second paper [23] on statistical aspects of molecular theory, included a theoretical analysis of the Brownian movement. Einstein wrote later of this: "My major aim in this was to find facts which would guarantee as much as possible the existence of atoms of definite size. In the midst of this I discovered that, according to atomistic theory, there would have to be a movement of suspended microscopic particles open to observation, without knowing that observations concerning the Brownian motion were already long familiar." * The third paper [13], on special relativity, included applications to electrodynamics such as the relativistic mass of a moving body, all subsequently confirmed experimentally.

Under these circumstances, it is not particularly fruitful to worry about whether, or to what extent, Einstein was aware of the Michelson-Morley experiment ** (the evidence is that he had heard of the result but not the details) or the directly relevant 1904 papers of Lorentz and Poincaré† (the evidence is strong that he had not read them)—all the more so since all the participants who have commented on it acknowledge Einstein as the original author of relativity theory. Instead, we should note another characteristic of Einstein's work, which

---

*Robert Brown, in 1827, had published these observations.
**See Refs. 24–26 for a fascinating analysis of this issue and of Einstein's early work.
† See Refs. 27 and 28 for a careful study of the historical situation and the characteristics of Einstein's work.

suggests why his approach to a problem was usually not that of the mainstream; namely, his attempt to restrict hypotheses to the smallest number possible and to the most general kind. For example, Lorentz, who never really accepted Einstein's relativity, used a great many *ad hoc* hypotheses to arrive at the same transformations in 1904 as Einstein did in 1905 (and as Voigt did in 1887); furthermore, Lorentz had assumed these equations *a priori* in order to obtain the invariance of Maxwell's equations in free space. Einstein on the other hand *derived* them from the simplest and most general postulates—the two fundamental principles of special relativity. And he was guided by his solution to the problem that had occupied his thinking since he was 16 years old: the nature of time. Lorentz and Poincaré had accepted Newton's universal time ($t = t'$), whereas Einstein abandoned that notion.

Newton, even more than many succeeding generations of scientists, was aware of the fundamental difficulties inherent in his formulation of mechanics, based as it was on the concepts of absolute space and absolute time. Einstein expressed a deep admiration for Newton's method and approach and can be regarded as bringing many of the same basic attitudes to bear on his analysis of the problem. In his Autobiographical Notes [18], after critically examining Newtonian mechanics, Einstein writes:

"Enough of this. Newton, forgive me; you found the only way which, in your age, was just about possible for a man of highest thought and creative power. The concepts, which you created, are even today still guiding our thinking in physics, although we now know that they will have to be replaced by others farther removed from the sphere of immediate experience, if we aim at a profounder understanding of relationships."

It seems altogether fitting that Einstein should have extended the range of Newton's relativity principle, generalized Newton's laws of motion, and later incorporated Newton's law of gravitation into his space-time scheme. In subsequent chapters we shall see how this was accomplished.

QUESTIONS

1. Can a particle move through a medium at a speed greater than the speed of light *in that medium?* Explain. (See R. Resnick and D. Halliday, *Physics,* p. 517–518.)

2. Is the sum of the interior angles of a triangle equal to 180° on a spherical surface? On a plane surface? Under what circumstances does spherical geometry reduce to plane geometry? Draw an analogy to relativistic mechanics and classical mechanics.

3. Would observers on the North Pole agree with those on the South Pole as to the direction of "up" and "down?" What definition of the terms could they agree on?

4. Give examples of non-inertial reference frames.

5. How does the concept of simultaneity enter into the measurement of the length of a body?

6. Could a mechanical experiment be performed in a given reference frame which would reveal information about the *acceleration* of that frame relative to an inertial one?

7. Discuss the following comment, which applies to most of the figures: "The figure *itself* belongs to some particular reference frame, that is, the picture represents measurements made in some particular frame." Can we look omnipotently at moving frames, wave fronts, and the like, without realizing first what frame *we* are in?

8. In an inelastic collision, the amount of thermal energy (internal mechanical kinetic energy) developed is independent of the inertial reference frame of the observer. Explain why, in words.

9. Describe an acoustic Michelson-Morley experiment by analogy with the optical one. What differences would you expect, and what similarities, in comparing the acoustical and the optical experiment?

10. Does the Lorentz-Fitzgerald contraction hypothesis contradict the classical notion of a rigid body?

11. Comment on the assertion (Ref. 29) that if one accepts Einstein's principle of the constancy of the velocity of light, there is then no reason to interpret the Michelson-Morley experiment as evidence against an ether hypothesis.

12. If the earth's motion, instead of being nearly circular about the sun, were uniformly along a straight line through the "ether," could an aberration experiment measure its speed?

13. How can we use the aberration observations to refute the Ptolemaic model of the solar system?

14. Does the fact that stellar aberration is observable contradict the principle of the relativity of uniform motion (i.e., does it determine an absolute velocity)? How, in this regard, does it differ from the Michelson-Morley experiment?

15. If the "ether" were dragged along with water, what would be the expected result of the aberration experiment when done with a telescope filled with water? (The actual results were the same with as without water. The experiment was done by Sir George Airy in 1871 and confirmed Eq. 1-12. For a complete analysis see Rosser [3].)

16. Of the various emission theories, only the original source theory is consistent with the ordinary optical result of the Doppler effect for a moving mirror. Explain.

17. What boxes in Table 1-2 have been accounted for in this chapter?

18. Of the experiments discussed in this chapter, which ones were not available at the time of Einstein's 1905 paper? (See references.)

PROBLEMS

1. Justify the relations $y = y'$ and $z = z'$ of Eq. 1-1a by symmetry arguments.

2. Momentum is conserved in a collision of two objects as measured by an observer on a uniformly moving train. Show that momentum is also conserved for a ground observer.

3. Repeat Problem 2 under the assumption that after the collision the masses of the two objects are different from what they were before; that is, assume a transfer of mass took place in the course of the collision. Show that for momentum to be conserved for the ground observer, conservation of mass must hold true.

4. Kinetic energy is conserved in an elastic collision by definition. Show, using the Galilean transformation equations, that if a collision is elastic in one inertial frame it is elastic in all inertial frames.

5. Consider two observers, one whose frame is attached to the ground and another whose frame is attached, say, to a train moving with uniform velocity **u** with respect to the ground. Each observes that a particle, initially at rest with respect to the train, is accelerated by a constant force applied to it for time t in the forward direction. (*a*) Show that for each observer the work done by the force is equal to the gain in kinetic energy of the particle, but that one observer measures these quantities to be $\frac{1}{2}ma^2t^2$, whereas the other observer measures them to be $\frac{1}{2}ma^2t^2 + maut$. Here a is the common acceleration of the particle of mass m. (*b*) Explain the differences in work done by the same force in terms of the different distances through which the observers measure the force to act during the time t. Explain the different final kinetic energies measured by each observer in terms of the work the particle could do in being brought to rest relative to each observer's frame.

★**6.** Suppose, in the previous problem, that there is friction between the particle and, say, the train floor, and that the applied force gives the particle the same acceleration over the same time as before. Note that there is no change in the initial and final kinetic energies but an extra force is needed to oppose friction. (*a*) Show that the amount of heat energy developed is the *same* for each observer. (*Hint.* Work done against friction depends on the *relative* motion of the surfaces.) (*b*) The applied force does work on the train itself, according to the ground observer, in addition to developing heat energy and increasing the kinetic energy of the particle. Compute the amount of this work. Is there an equivalent performance of work by the observer on the train? Explain.

7. Write the Galilean transformation equations for the case of arbitrary relative velocity of the frames. (*Hint.* Let **v** have components v_x, v_y, and v_z.)

8. A pilot is supposed to fly due east from A to B and then back again to A due west. The velocity of the plane in air is u' and the velocity of the air with respect to the ground is v. The distance between A and B is l and the plane's air speed u' is constant. (*a*) If $v = 0$ (still air), show that the time for the round trip is $t_0 = 2l/u'$. (*b*) Suppose that the air velocity is due east (or west). Show that the time for a round trip is then

$$t_E = \frac{t_0}{1 - v^2/(u')^2}.$$

(*c*) Suppose that the air velocity is due north (or south). Show that the time for a round trip is then

$$t_N = \frac{t_0}{\sqrt{1 - v^2/(u')^2}}.$$

(*d*) In parts (*b*) and (*c*) we must assume that $v < u'$. Why? (*e*) Draw an analogy to the Michelson-Morley experiment.

9. One example of the binomial expansion, in which a function of x is expanded into a series of terms of higher and higher order in x, is

$$\frac{1}{1 - x} = 1 + x + x^2 + x^3 + \cdots.$$

(*a*) Prove this result by direct division of $1 - x$ into 1. (*b*) Then show that if $x \ll 1$ one can set $1/(1 - x) \approx 1 + x$. Consider specifically the cases in which $x = 0.01$ and $x = 0.10$. (*c*) Show that if $x \approx 1$ the approximation of part (*b*) is very poor. Consider specifically the cases in which $x = 0.6$ and $x = 0.9$.

★**10.** In the description of the Michelson-Morley experiment, it was assumed that one of the arms of the interferometer was aligned along the direction of the earth's motion while the second was perpendicular to this direction. Suppose, instead, that one arm makes an angle of ϕ with the direction of motion (see Fig. 1-11). Repeat the analysis in the text for this more general case and show that, under the Lorentz-Fitzgerald contraction hypothesis, no fringe shift would be expected when the apparatus is rotated through 90°; that is, the time difference between the two beams is the same before and after rotation. (*Hint.* Remember, only the component of length in the direction of motion through the ether is affected).

11. Derive Eq. 1-10.

★**12.** Using the plane defined in Fig. 1-9*b*, show that, to first order in v/c, Eq. 1-11 becomes $\tan \alpha = (v/c) \sin \theta$ when the rays from a star make an arbitrary angle θ with the plane of the earth's orbit, rather than an angle $\theta = 90°$ as assumed for simplicity in the text. (That is, **v** and **c** are no longer at right angles and angles α and θ are in the same plane.) Does this change the conclusions drawn there?

13. (*a*) In the Fizeau experiment (Fig. 1-10) identify the frames S and S' and the relative velocity **v** which correspond to Fig. 1-1. (*b*) Show that in the Fresnel drag formula (Eq. 1-12) $v \rightarrow \frac{c}{n} + v_w \rightarrow v_w$ for very large values of n. How would you interpret this? (*c*) Under what circumstances will the Fresnel drag coefficient be zero? To what does this correspond physically?

14. Consider one star in a binary system moving in uniform circular motion with speed v. Consider two positions: (I) the star is moving *away* from the earth along the line connecting them, and (II) the star is moving *toward* the earth along the line connecting them (see Fig. 1-12). Let the period of the star's motion be T and its distance from earth be l. Assume l is large enough that positions I and II are a

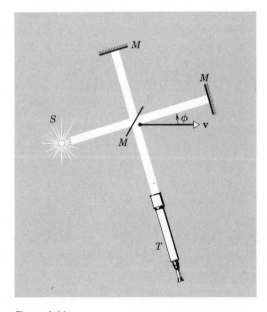

Figure 1-11.

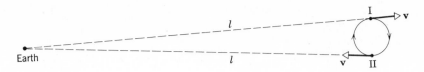

Figure 1-12.

half-orbit apart. (*a*) Show that the star would appear to go from position (I) to position (II) in a time $T/2 - 2lv/(c^2 - v^2)$ and from position (II) to position (I) in a time $T/2 + 2lv/(c^2 - v^2)$, assuming that the emission theories are correct. (*b*) Show that the star would appear to be at both positions I and II at the same time if $T/2 = 2lv/(c^2 - v^2)$.

15. A bullet from a rifle travels 1100 ft in its first second of motion. On a calm day the rifle is fired from a train along the tracks. A man stands 1100 ft away from the rifle at that instant, in the line of fire. Does the bullet or the sound of the firing reach the man first if the train (*a*) is at rest, (*b*) is moving away from the man, or (*c*) is moving toward the man? (*d*) Is the first sentence of this problem ambiguous? Explain. (*e*) State the relevance of this problem to emission theories.

REFERENCES

1. See the film "The Ultimate Speed" by William Bertozzi (produced by Educational Services, Inc., Watertown, Mass.) and a complete description of it by W. Bertozzi in *Am. J. Phys.*, **32**, 551–555 (1964).

2. W. K. H. Panofsky and Melba Phillips, *Classical Electricity and Magnetism*, (Addison-Wesley, Reading, Mass., 1955) Chapter 14.

3. F. T. Trouton and H. R. Noble, *Phil. Trans. Roy. Soc.*, **A 202**, 165 (1903); *Proc. Roy. Soc. (London)*, **72**, 132 (1903). A concise account is given in W. G. V. Rosser, *An Introduction to the Theory of Relativity* (Butterworths, London, 1964), pp. 64–65. This text is an outstanding general reference.

4. A. A. Michelson, *Am. J. Sci.*, **122**, 120 (1881). A. A. Michelson and E. W. Morley, *Am. J. Sci.*, **134**, 333 (1887).

5. J. P. Cedarholm, G. L. Bland, B. L. Havens, and C. H. Townes, "New Experimental Tests of Special Relativity," *Phys. Rev. Letters*, **1**, 342–343 (1958).

6. R. J. Kennedy and E. M. Thorndike, *Phys. Rev.*, **42**, 400 (1932).

7. See Albert Stewart, "The Discovery of Stellar Aberration," *Scientific American*, p. 100 (March 1964) for an interesting and detailed description of Bradley's work.

8. The original source emission theory is that of W. Ritz, *Ann. Chim. et Phys.*, **13**, 145 (1908). Discussions of various emission theories can be found in R. C. Tolman, *Phys. Rev.*, **31**, 26 (1910); J. J. Thomson, *Phil. Mag.*, **19**, 301 (1910; and Stewart, *Phys. Rev.*, **32**, 418 (1911).

9. W. De Sitter, *Proc. Amsterdam Acad.*, **15**, 1297 (1913), and **16**, 395 (1913).

10. "Evidence Against Emission Theories," J. G. Fox, *Am. J. Phys.*, **33**, 1 (1965).

11. D. Sadeh, *Phys. Rev. Letters*, **10**, 271 (1963) (a measurement of the speed of electromagnetic radiation from the annihilation of rapidly moving positrons); "Test of the Second Postulate of Special Relativity in the GeV Region" by T. Alväger, F. J. M. Farley, J. Kjellman, and I. Wallin, *Phys. Letters*, **12**, 260 (1964).

12. R. Tomaschek, *Ann. Phys. (Leipzig)*, **73**, 105 (1924). D. C. Miller, *Proc. Nat. Acad. Sci.*, **2**, 311 (1925).

13. A. Einstein, "On the Electrodynamics of Moving Bodies," *Ann. Physik*, **17**, 891 (1905).

14. For a translated extract see "Great Experiments in Physics," edited by Morris H. Shamos (Holt Dryden, New York, 1959), p. 318.

15. H. Bondi, *Endeavour,* **20,** 121 (1961).

16. "Einstein and Some Civilized Discontents," by Martin J. Klein, *Physics Today,* **18,** No. 1, 38 (1965).

17. Barbara Lovett Cline, *The Questioners* (Crowell, New York, 1965). See in this connection, Chapters 5 and 12.

18. P. A. Schlipp (ed.), *Albert Einstein: Philosopher-Scientist* (Harper Torchbooks, New York, 1959), a two-volume work.

19. Elma Ehrlich Levinger, *Albert Einstein* (Julian Messner, New York, 1949).

20. Peter Michelmore, *Einstein, Profile of the Man* (Dodd, Mead and Co., New York, 1962). In paperback, Apollo Editions, A-63.

21. William Cahn, *Einstein, A Pictorial Biography* (The Citadel Press, New York, 1955). In paperback, 1960.

22. A. Einstein, *Ann. Physik,* **17,** 132 (1905).

23. A. Einstein, *Ann. Physik,* **17,** 549 (1905).

24. R. S. Shankland, "Conversations with Albert Einstein," *Am. J. Phys.,* **31,** 47 (1963).

25. R. S. Shankland, "Michelson-Morley Experiment," *Am. J. Phys.,* **32,** 16 (1964).

26. Gerald Holton, "Einstein and the 'Crucial' Experiment," *Am. J. Phys.,* **37,** 968 (1969).

27. Gerald Holton, "On the Origins of the Special Theory of Relativity," *Am. J. Phys.,* **28,** 627 (1960).

28. W. Rindler, "Einstein's Priority in Recognizing Time Dilation Physically," *Am. J. Phys.,* **38,** 1111 (1970).

29. Håken Törnebohm, "Two Studies Concerning the Michelson-Morley Experiment," Foundations of Physics, Vol. 1, No. 1, p. 47 (1970).

CHAPTER 2

Relativistic Kinematics

2.1 The Relativity of Simultaneity

In *Conversations with Albert Einstein*, R. S. Shankland [1] writes "I asked Professor Einstein how long he had worked on the Special Theory of Relativity before 1905. He told me that he had started at age 16 and worked for ten years; first as a student when, of course, he could spend only part-time on it, but the problem was always with him. He abandoned many fruitless attempts, 'until at last it came to me that time was suspect!'" What was it about time that Einstein questioned? It was the assumption, often made unconsciously and certainly not stressed, that there exists a universal time which is the same for all observers. Indeed, it was only to bring out this assumption explicitly that we included the equation $t = t'$ in the Galilean transformation equations (Eq. 1-1). In pre-relativistic discussions, the assumption was there implicitly by the absence of a transformation equation for t in the Galilean equations. That the same time scale applied to all inertial frames of reference was a basic premise of Newtonian mechanics.

In order to set up a universal time scale, we must be able to give meaning, independent of a frame of reference, to statements such as "Events A and B occurred at the same time." Einstein pointed out that when we say that a train arrives at 7 o'clock this means that the exact pointing of the clock hand to 7 and the arrival of the train at the clock were simultaneous. We certainly shall not have a universal time scale if different inertial observers disagree as to whether two events are simultaneous. Let us first try to set up an unambiguous time scale in a single frame of reference; then we can set up time scales in exactly the same way in all inertial frames and compare what different observers have to say about the sequence of two events, A and B.

Suppose that the events occur at the same place in one particular frame of reference. We can have a clock at that place which registers the time of occurrence of each event. If the reading is the same for each event, we can logically regard the events as simultaneous. But what if the two events occur at *different* locations? Imagine now that there is a clock at the positions of each event—the clock at A being of the same nature as that at B, of course. These clocks can record the time of occurrence of the events but, before we can compare their readings, we must be sure that they are synchronized.

Some "obvious" methods of synchronizing clocks turn out to be erroneous. For example, we can set the two clocks so that they always read the same time *as seen by* observer A. This means that whenever A looks at the B clock it reads the same to him as his clock. The defect here is that if observer B uses the same criterion (that is, that the clocks are synchronized if they always read the same time to *him*), he will find that the clocks are *not* synchronized if A says that they *are*. The reason is that this method neglects the fact that it takes time for light to travel from B to A and vice versa. The student should be able to show that, if the distance between the clocks is L, one observer will see the other clock lag his by $2L/c$ when the other observer claims that they are synchronous. We certainly cannot have observers in the same reference frame disagree on whether clocks are synchronized or not, so we reject this method.

An apparent way out of this difficulty is simply to set the two clocks to read the same time and then move them to the positions where the events occur. (In principle, we need clocks everywhere in our reference frame to record the time of occurrence of events, but once we know how to synchronize two clocks we can, one by one, synchronize all the clocks.) The difficulty here is that we do not know ahead of time, and therefore cannot assume, that the motion of the clocks (which may have different velocities, accelerations, and path lengths in being moved into position) will not affect their readings or time-keeping ability. Even in classical physics, the motion can affect the rate at which clocks run.

Hence, the logical thing to do is to put our clocks into position and synchronize them by means of signals. If we had a method of transmitting signals with infinite speed, there would be no complications. The signals would go from clock A to clock B to clock C, and so on, in zero time. We could use such a signal to set all clocks at the same time reading. But no signal known has this property. All known signals require a finite time to travel some distance, the time increasing with the distance traveled. The best signal to choose would be one whose speed depends on as few factors as possible. We choose electromagnetic waves because they do not require a material medium for transmission and their speed in vacuum does not depend on their wavelength, amplitude, or direction of propagation. Furthermore, their propagation speed is the highest known and—most important for finding a universal method of synchronization—experiment shows their speed to be the same for all inertial observers.

Now we must account for the finite time of transmission of the signal and our clocks can be synchronized. To do this let us imagine an observer with a light source that can be turned on and off (e.g., a flash bulb) at each clock, A and B. Let the measured distance between the clocks (and observers) be L. The agreed-upon procedure for synchronization then is that A will turn on his light source when his clock reads $t = 0$ and observer B will set his clock to $t = L/c$ the instant he receives the signal. This accounts for the transmission time and synchronizes the clocks in a consistent way. For example, if B turns on his light source at some later time t by his clock, the signal will arrive at A at a time $t + L/c$, which is just what A's clock will read when A receives the signal.

A method equivalent to the above is to put a light source at the exact midpoint of the straight line connecting A and B and inform each observer to put his clock at $t = 0$ when the turned-on light signal reaches him. The light will take an equal amount of time to reach A and B from the midpoint, so that this procedure does indeed synchronize the clocks.

Now that we have a procedure for synchronizing clocks in one reference frame, we can judge the time order of events in that frame. The time of an event is measured by the clock whose location coincides with that of the event. Events occurring at two different places in that frame must be called *simultaneous* when the clocks at the respective places record the same time for them. Suppose that one inertial observer does find that two separated events are simultaneous. Will these same events be measured as simultaneous by an observer on another inertial frame which is moving with speed v with respect to the first? (Remember, each observer uses an identical procedure to synchronize the clocks in his reference frame.) If not, simultaneity is not independent of the frame of reference used to describe events. Instead of being absolute, simultaneity would be a relative concept. Indeed, this is exactly what we find to be true, in direct contradiction to the classical assumption.

To understand this, let us consider an example. Let there be two inertial reference frames S' and S having a relative velocity. Each frame has its own meter sticks and synchronized clocks. The observers note that two lightning bolts strike each, hitting and leaving permanent marks in the frames.* Assume that afterwards, by measurements, each inertial observer finds that he was located exactly at the midpoint of the marks which were left on his reference frame. In Fig. 2-1a, these marks are left at A and B on the S-frame and at A' and B' on the S' frame, the observers being at O and O'. Because each observer knows he was at the midpoint of the mark left by these events, he will conclude that they were simultaneous if the light signals from them arrive simultaneously at his clock (see the definitions of simultaneity given earlier). If, on the other hand, one signal arrives before the other, he will conclude that one event preceded the other. Since each observer has a synchronized set of clocks, he can conclude either that the clocks at the marks read the same time when the marks were made (simultaneous case) or that they read different times (non-simultaneous case).

Many different possibilities exist in principle as to what the measurements might show. Let us suppose, for the sake of argument, that the S-observer finds that the lightning bolts struck simultaneously. Will the S'-observer also find these events to be simultaneous? In Figs. 2-1b to 2-1d we take the point of view of the S-observer and see the S'-frame moving, say, to the right. At the instant the lightning struck at A and A', these two points coincide, and

---

*The essential point is to have light sources that leave marks. Exploding sticks of dynamite would do as well, for example.

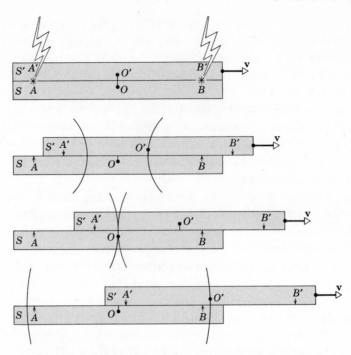

Figure 2-1. *The point of view of the S-frame,* the *S'*-frame moving to the right. A light wave leaves *A, A'* and *B, B'* in (*a*). Successive drawings correspond to the assumption that event *AA'* and event *BB'* are simultaneous in the *S*-frame. In (*b*) one wavefront reaches *O'*. In (*c*) both wavefronts reach *O*. In (*d*) the other wavefront reaches *O'*.

at the instant the lightning struck at *B* and *B'* those two points coincide. The *S*-observer found these two events to occur at the same instant, so that at that instant *O* and *O'* must coincide also for him. However, *the light signals from the events take a finite time to reach O and during this time O' travels to the right* (Figs. 2-1*b* to 2-1*d*). Hence, the signal from event *BB'* arrives at *O'* (Fig. 2-1*b*) before it gets to *O* (Fig. 2-1*c*), whereas the signal from event *AA'* arrives at *O* (Fig. 2-1*c*) before it gets to *O'* (Fig. 2-1*d*). Consistent with our starting assumption, the *S*-observer finds the events to be simultaneous (both signals arrive at *O* at the same instant). The *S'*-observer, however, finds that event *BB'* precedes event *AA'* in time; they are *not* simultaneous to him. Therefore, two separated events which are simultaneous with respect to one frame of reference are not necessarily simultaneous with respect to another frame.

Now we could have supposed, just as well, that the lightning bolts struck so that the *S'*-observer found them to be simultaneous. In that case the light signals reach *O'* simultaneously, rather than *O*. We show this in Fig. 2-2 where now we take the point of view of *S'*. The *S*-frame moves to the left relative to the *S'*-observer. But, in this case, the signals do not reach *O* simultaneously; the signal from event *AA'* reaches *O* before that from event *BB'*. Here the *S'*-observer finds the events to be simultaneous but the *S*-observer finds that event *AA'* precedes event *BB'*.

Hence, *neither* frame is preferred and the situation is perfectly reciprocal. Simultaneity is genuinely a relative concept, not an absolute one.* Indeed, the two figures become indistinguishable if you turn one of them upside down. Neither observer can assert absolutely that he is at rest. Instead, each observer correctly states only that the other one is moving relative to him and that the signals travel with finite speed c relative to him. It should be clear that if we had an infinitely fast signal, then simultaneity *would* be an absolute concept; for the frames would not move at all relative to one another in the (zero) time it would take the signal to reach the observers.

Some other conclusions suggest themselves from the relativity of simultaneity. To measure the length of an object means to locate its end points simultaneously. Because simultaneity is a relative concept, length measurements will also depend on the reference frame and be relative. Furthermore, we find that the rates at which clocks run also depend on the reference frame. This can be illustrated as follows. Consider two clocks, one on a train and

---

*The time order of two events *at the same place* can be absolutely determined. It is in the case that two events are *separated* in space that simultaneity is a relative concept. In our arguments, we have shown that *if* one observer finds the events to be simultaneous, *then* the other one will find them not to be simultaneous. Of course, it could also happen that neither observer finds the events to be simultaneous but then they would disagree either on the time order of the events or on the time interval elapsing between the events, or both (see Ref. 2).

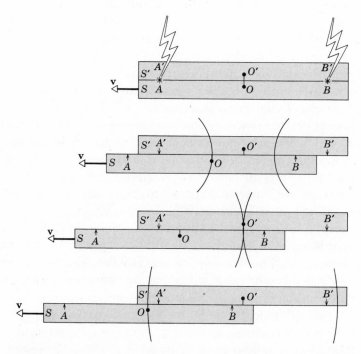

Figure 2-2. *The point of view of the S′-frame,* the S-frame moving to the left. A light wave leaves A, A' and B, B' in (a). Successive drawings correspond to the assumption that event AA' and event BB' are simultaneous in S′-frame. In (b) one wavefront reaches O. In (c) both wavefronts reach O'. In (d) the other wavefront reaches O.

one on the ground, and assume that at the moment they pass one another (i.e., the instant that they are coincident) they read the same time (i.e., the hands of the clocks are in identical positions). Now, if the clocks continue to agree, we can say that they go at the same rate. But, when they are a great distance apart, we know from the preceding discussion that their hands cannot have identical positions simultaneously as measured both by the ground observer and the train observer. Hence, time interval measurements are also relative, that is, they depend on the reference frame of the observer. As a result of the relativity of length and time interval measurements it is perhaps possible to reconcile ourselves to the experimental fact that observers who are moving relative to each other measure the speed of light to be the same. In succeeding sections, we shall look more carefully into these matters.

2.2 Derivation of the Lorentz Transformation Equations

We have seen that the Galilean transformation equations must be replaced by new ones consistent with experiment. Here we shall derive these new equations, using the postulates of special relativity theory. To show the consistency of the theory with the discussion of the previous section, we shall then derive all the special features of the new transformation equations again from the more physical approach of the measurement processes discussed there.

We observe an event in one inertial reference frame S and characterize its location and time by specifying the coordinates x, y, z, t of the event. In a second inertial frame S', this *same event* is recorded as the space-time coordinates x', y', z', t'. We now seek the functional relationships $x' = x'(x,y,z,t)$, $y' = y'(x,y,z,t)$, $z' = z'(x,y,z,t)$, and $t' = t'(x,y,z,t)$. That is, we want the equations of transformation which relate one observer's space-time coordinates of an event with the other observer's coordinates of the same event.

We shall use the fundamental postulates of relativity theory and, in addition, the assumption that space and time are homogeneous. This homogeneity assumption (which can be paraphrased by saying that all points in space and time are equivalent) means, for example, that the results of a measurement of a length or time interval between two specific events should not depend on where or when the interval happens to be in our reference frame. We shall illustrate its application shortly.

We can simplify the algebra by choosing the relative velocity of the S and S' frames to be along a common x-x' axis and by keeping corresponding planes parallel (see Fig. 1-1). This does not impose any fundamental restrictions on our results for space is isotropic (that is, has the same properties in all directions) a result contained in the homogeneity assumption. Also, at the instant the origins O and O' coincide, we let the clocks there read $t = 0$ and $t' = 0$, respectively. Now, as explained below, the homogeneity assumption requires that transformation equations must be linear (i.e., they involve only the first power in the variables), so that the most general form they can take is

$$\begin{aligned} x' &= a_{11}x + a_{12}y + a_{13}z + a_{14}t \\ y' &= a_{21}x + a_{22}y + a_{23}z + a_{24}t \\ z' &= a_{31}x + a_{32}y + a_{33}z + a_{34}t \\ t' &= a_{41}x + a_{42}y + a_{43}z + a_{44}t. \end{aligned} \tag{2-1}$$

Here, the subscripted coefficients are constants that we must determine to obtain the exact transformation equations. Notice that we do not exclude the possible dependence of space and time coordinates upon one another.

If the equations were not linear, we would violate the homogeneity assumption. For example, suppose that x' depended on the square of x, that is, as $x' = a_{11}x^2$. Then the distance between two points in the primed frame would be related to the location of these points in the unprimed frame by $x_2' - x_1' = a_{11}(x_2^2 - x_1^2)$. Suppose now that a rod of unit length in S had its end points at $x_2 = 2$ and $x_1 = 1$; then $x_2' - x_1' = 3a_{11}$. If, instead, the same rod happens to be located at $x_2 = 5$ and $x_1 = 4$, we would obtain $x_2' - x_1' = 9a_{11}$. That is, the measured length of the rod would depend on where it is in space. Likewise, we can reject any dependence on t that is not linear, for the time interval between two events should not depend on the numerical setting of the hands of the observer's clock. The relationships must be linear then in order not to give the choice of origin of our space-time coordinates (or some other point) a physical preference over all other points.

Now, regarding these sixteen coefficients, it is expected that their values will depend on the relative velocity v of the two inertial frames. For example, if $v = 0$, then the two frames coincide at all times and we expect $a_{11} = a_{22} = a_{33} = a_{44} = 1$, all other coefficients being zero. More generally, if v is small compared to c, the coefficients should lead to the (classical) Galilean transformation equations. We seek to find the coefficients for *any* value of v, that is, as functions of v.

How then do we determine the values of these sixteen coefficients? Basically, we use the postulates of relativity, namely (1) The Principle of Relativity—that no preferred inertial system exists, the laws of physics being the same in all inertial systems—and (2) The Principle of the Constancy of the Speed of Light—that the speed of light in free space has the same value c in all inertial systems. Let us proceed.

With no relative motion of the frames in the y- or z-direction, we might expect $y' = y$ and $z' = z$. This result does indeed follow directly from arguments using the relativity postulate (see Sec. 2-2 of Ref. 2 for a proof), so that $a_{22} = 1 = a_{33}$ and $a_{21} = a_{23} = a_{24} = a_{31} = a_{32} = a_{34} = 0$ and eight of the coefficients are thereby determined. Therefore, our two middle transformation equations become

$$y' = y \quad \text{and} \quad z' = z. \tag{2-2}$$

There remain transformation equations for x' and t', namely,

$$x' = a_{11}x + a_{12}y + a_{13}z + a_{14}t$$
and
$$t' = a_{41}x + a_{42}y + a_{43}z + a_{44}t.$$

Let us look first at the t'-equation. For reasons of symmetry, we assume that t' does not depend on y and z. Otherwise, clocks placed symmetrically in the y-z plane (such as at $+y$, $-y$ or $+z$, $-z$) about the x-axis would appear to disagree as observed from S', which would contradict the isotropy of space. Hence, $a_{42} = a_{43} = 0$. As for the x'-equation, we know that a point having $x' = 0$ appears to move in the direction of the positive x-axis with speed v, so that the statement $x' = 0$ must be identical to the statement $x = vt$. Therefore, we expect $x' = a_{11}(x - vt)$ to be the correct transformation equation. (That is, $x = vt$ always gives $x' = 0$ in this equation.) Hence, $x' = a_{11}x - a_{11}vt = a_{11}x + a_{14}t$. This gives us $a_{14} = -va_{11}$, and our four equations have now been reduced to

$$
\begin{aligned}
x' &= a_{11}(x - vt) \\
y' &= y \\
z' &= z \\
t' &= a_{41}x + a_{44}t.
\end{aligned}
\tag{2-3}
$$

There remains the task of determining the three coefficients a_{11}, a_{41}, and a_{44}. To do this, we use the principle of the constancy of the velocity of light. Let us assume that at the time $t = 0$ a spherical electromagnetic wave leaves the origin of S, which coincides with the origin of S' at that moment. The wave propagates with a speed c in all directions in each inertial frame. Its progress, then, is described by the equation of sphere whose radius expands with time at the same rate c in terms of *either* the primed or unprimed set of coordinates. That is,

$$
x^2 + y^2 + z^2 = c^2 t^2
\tag{2-4}
$$

or

$$
x'^2 + y'^2 + z'^2 = c^2 t'^2.
\tag{2-5}
$$

If now we substitute into Eq. 2-5 the transformation equations (Eqs. 2-3), we get

$$
a_{11}^2(x - vt)^2 + y^2 + z^2 = c^2(a_{41}x + a_{44}t)^2.
$$

Rearranging the terms gives us

$$
(a_{11}^2 - c^2 a_{41}^2)x^2 + y^2 + z^2 - 2(va_{11}^2 + c^2 a_{41}a_{44})xt = (c^2 a_{44}^2 - v^2 a_{11}^2)t^2.
$$

In order for this expression to agree with Eq. 2-4, which represents the same thing, we must have

$$
\begin{aligned}
c^2 a_{44}^2 - v^2 a_{11}^2 &= c^2 \\
a_{11}^2 - c^2 a_{41}^2 &= 1 \\
va_{11}^2 + c^2 a_{41}a_{44} &= 0.
\end{aligned}
$$

Here we have three equations in three unknowns, whose solution (as the student can verify by substitution into the three equations above) is

$$
\begin{aligned}
a_{44} &= 1/\sqrt{1 - v^2/c^2} \\
a_{11} &= 1/\sqrt{1 - v^2/c^2}
\end{aligned}
\tag{2-6}
$$

and

$$
a_{41} = -\frac{v}{c^2}/\sqrt{1 - v^2/c^2}.
$$

By substituting these values into Eqs. 2-3, we obtain, finally, the new sought-after transformation equations,

$$
\begin{aligned}
x' &= \frac{x - vt}{\sqrt{1 - v^2/c^2}} \\
y' &= y \\
z' &= z \\
t' &= \frac{t - (v/c^2)x}{\sqrt{1 - v^2/c^2}},
\end{aligned}
\tag{2-7}
$$

the so-called* *Lorentz transformation equations.*

Before probing the meaning of these equations, we should put them to two necessary tests. First, if we were to exchange our frames of reference or—what amounts to the same thing—consider the given space-time coordinates of the event to be those observed in S' rather than in S, the only change allowed by the relativity principle is the physical one of a change in relative velocity from v to $-v$. That is, from S' the S-frame moves to the left whereas from S the S'-frame moves to the right. When we solve Eqs. 2-7 for x, y, z, and t in terms of the primed coordinates (see Problem 3), we obtain

$$x = \frac{x' + vt'}{\sqrt{1 - v^2/c^2}},$$
$$y = y',$$
$$z = z', \tag{2-8}$$
$$t = \frac{t' + (v/c^2)x'}{\sqrt{1 - v^2/c^2}}$$

which are identical in form with Eqs. 2-7 except that, as required, v changes to $-v$.

Another requirement is that for speeds small compared to c, that is, for $v/c \ll 1$, the Lorentz equations should reduce to the (approximately) correct Galilean transformation equations. This is the case, for when $v/c \ll 1$, Eqs. 2-7 become**

$$\begin{aligned} x' &= x - vt \\ y' &= y \\ z' &= z \\ t' &= t \end{aligned} \tag{2-9}$$

which are the classical Galilean transformation equations.

In Table 2-1 we summarize the Lorentz transformation equations.

2.3 Some Consequences of the Lorentz Transformation Equations

The Lorentz transformation equations (Eqs. 2-7 and 2-8), derived rather formally in the last section from the relativity postulates, have some interesting consequences for length and time measurements. We shall look at them briefly in this section. In the next section we shall present a more physical interpretation of these equations and their consequences, relating them directly to the operations of physical measurement. Throughout the chapter we shall cite experiments that confirm these consequences.

---

*Poincaré originally gave this name to the equations. Lorentz, in his classical theory of electrons, had proposed them before Einstein did. However, Lorentz took v to be the speed relative to an absolute ether frame and gave a different interpretation to the equations.

**In the time equation, $t' = (t - vx/c^2)/\sqrt{1 - v^2/c^2}$, consider the motion of the origin O', for example, given by $x = vt$. Then

$$t' = (t - v^2t/c^2)/\sqrt{1 - v^2/c^2} = t\sqrt{1 - v^2/c^2}.$$

As $v/c \rightarrow 0$, $t' \rightarrow t$.

Table 2-1 THE LORENTZ TRANSFORMATION EQUATIONS

$x' = \dfrac{x - vt}{\sqrt{1 - v^2/c^2}}$	$x = \dfrac{x' + vt'}{\sqrt{1 - v^2/c^2}}$
$y' = y$	$y = y'$
$z' = z$	$z = z'$
$t' = \dfrac{t - (v/c^2)x}{\sqrt{1 - v^2/c^2}}$	$t = \dfrac{t' + (v/c^2)x'}{\sqrt{1 - v^2/c^2}}$

One consequence is this: *a body's length is measured to be greatest when the body is at rest relative to the observer. When it moves with a velocity v relative to the observer its measured length is contracted in the direction of its motion by the factor $\sqrt{1 - v^2/c^2}$, whereas its dimensions perpendicular to the direction of motion are unaffected.* To prove the italicized statement, imagine a rod lying at rest along the x'-axis of the S'-frame. Its end points are measured to be at x_2' and x_1', so that its rest length is $x_2' - x_1'$. What is the rod's length as measured by the S-frame observer, for whom the rod moves with the relative speed v? For convenience, we shall let $v/c = \beta$, as before. From the first Lorentz equation we have

$$x_2' = \frac{x_2 - vt_2}{\sqrt{1 - \beta^2}} \qquad x_1' = \frac{x_1 - vt_1}{\sqrt{1 - \beta^2}}$$

so that

$$x_2' - x_1' = \frac{(x_2 - x_1) - v(t_2 - t_1)}{\sqrt{1 - \beta^2}}$$

Now the length of the rod in the S-frame is simply the distance between the end points, x_2 and x_1, of the moving rod measured at the same instant in that frame. Hence, with $t_2 = t_1$, we obtain

$$x_2' - x_1' = \frac{x_2 - x_1}{\sqrt{1 - \beta^2}}$$

or

$$x_2 - x_1 = (x_2' - x_1')\sqrt{1 - \beta^2} \qquad\qquad (2\text{-}10)$$

so that the measured length of the moving rod, $x_2 - x_1$, is contracted by the factor $\sqrt{1 - \beta^2}$ from its rest length, $x_2' - x_1'$. As for the dimensions of the rod along y and z, perpendicular to the relative motion, it follows at once from the transformation equations $y' = y$ and $z' = z$ that these are measured to be the same by both observers.

A second consequence is this: *A clock is measured to go at its fastest rate when it is at rest relative to the observer. When it moves with a velocity v relative to the observer, its rate is measured to have slowed down by a factor $\sqrt{1 - v^2/c^2}$.* To prove these italicized statements, consider a clock to be at rest at the position x' in the S'-frame. It may simplify matters to picture the hand of this clock going around and to let unit time be the time it takes the hand of the clock to go around once. Hence, the events we observe (the two successive coincidences of the hand of the clock with a given marker on the face of the clock) span the time interval $t_1' = t'$ to $t_2' = t' + 1$ in the primed coordinates. The S-frame observer records these events as occurring at times

$$t_1 = \frac{t_1' + (v/c^2)x_1'}{\sqrt{1 - \beta^2}} \quad \text{and} \quad t_2 = \frac{t_2' + (v/c^2)x_2'}{\sqrt{1 - \beta^2}}$$

which, with $t_1' = t'$ and $t_2' = t' + 1$ and with x_2' and x_1' each being the rest position x', become

$$t_1 = \frac{t' + (v/c^2)x'}{\sqrt{1 - \beta^2}} \quad \text{and} \quad t_2 = \frac{(t' + 1) + (v/c^2)x'}{\sqrt{1 - \beta^2}}$$

The clock in the S'-frame is at a fixed position x', but the times t_1 and t_2 are read from two *different* synchronized clocks in the S-frame, namely the stationary S-clock that happens to be coincident with the moving clock at the beginning of the interval, and the stationary S-clock coincident with the moving clock at the end of the interval. The time interval they record for the event is simply, from the above equations,

$$t_2 - t_1 = \frac{1}{\sqrt{1 - \beta^2}}.$$

Hence, unit time measured on the S'-clock is recorded as a *longer* time on the S-clocks. Clearly, if, instead of unit time, the S'-clock recorded a time interval $t_2' - t_1'$, the S-clock would have recorded the corresponding interval

$$t_2 - t_1 = \frac{t_2' - t_1'}{\sqrt{1 - \beta^2}}. \tag{2-11}$$

A time interval measured on the S'-clock is recorded as a *longer* time interval on the S-clocks (see Fig. 2-3). From the point of view of observer S, the moving S'-clock appears slowed down, that is, it appears to run at a rate which is slow by the factor $\sqrt{1 - \beta^2}$. This result applies to all S'-clocks observed from S, for the location x' in our proof was arbitrary.

It is common in relativity to speak of the frame in which the observed body is at rest as the *proper frame*. The length of a rod in such a frame is then called the *proper length*. Likewise, the *proper time interval* is the time interval recorded by a clock attached to the observed body. The proper time interval can be thought of equivalently as the time interval between two events occurring at the same place in the S'-frame or the time interval measured by a single clock at one place. A nonproper (or improper) time interval would be a time interval measured by two different clocks at two different places. Thus, we see from the previous discussion that if $d\tau$ represents a proper time interval, then the expression

$$dt = \frac{d\tau}{\sqrt{1 - \beta^2}} \tag{2-12}$$

relates the nonproper interval dt to the proper interval $d\tau$. Later we shall define other proper quantities, such as proper mass, and shall find that they represent invariant quantities in relativity theory.

A third consequence of the Lorentz-transformation equations is this: Although clocks in a moving frame all appear to go at the same slow rate when observed from a stationary frame with respect to which the clocks move, the

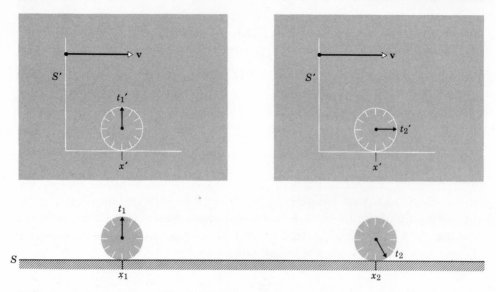

Figure 2-3. A clock at rest at x' in the S'-frame moves with a velocity v relative to the S-frame. The time interval this *one moving clock* records between passing two stationary (synchronized) clocks in $t_2' - t_1'$; these separate events are recorded in S as occurring at t_1 and at t_2 by the *two stationary clocks*. The recorded time interval $t_2 - t_1$ is *greater* than the time interval $t_2' - t_1'$. The moving clock appears to run slow.

moving clocks appear to differ from one another in their readings by a phase constant which depends on their location, that is, *they appear to be unsynchronized.* This becomes evident at once from the transformation equation

$$t = \frac{t' + (v/c^2)x'}{\sqrt{1 - \beta^2}}.$$

For consider an instant of time in the S-frame, that is, a given value of t. Then, to satisfy this equation, $t' + (v/c^2)x'$ must have a definite fixed value. This means the greater is x' (i.e., the farther away an S'-clock is stationed on the x'-axis) the smaller is t' (i.e., the farther behind in time its reading appears to be). Hence, the moving clocks appear to be out of phase, or synchronization, with one another. We shall see in the next section that this is just another manifestation of the fact that two events that occur simultaneously in the S-frame are not, in general, measured to be simultaneous in the S'-frame, and vice versa.

All the results of this section are reciprocal. That is, no matter which frame we take as the proper frame, the observer in the other frame measures a contracted length and dilated (expanded) time interval and finds the moving clocks to be out of synchronization.

▶ **Example 1.** The factor $\sqrt{1 - \beta^2}$ occurs in Eq. 2-10 and the factor $\gamma = 1/\sqrt{1 - \beta^2}$ in Eq. 2-12. Because they arise frequently in relativity, it is helpful to be able to estimate their values as a function of β. Compute $\sqrt{1 - \beta^2}$ and $\gamma = 1/\sqrt{1 - \beta^2}$ for $\beta = v/c = 0.100, 0.300, 0.600, 0.800, 0.900, 0.950$, and 0.990, and plot them as functions of β.

We find

$\beta =$	0.100	0.300	0.600	0.800	0.900	0.950	0.990
$\sqrt{1-\beta^2} =$	0.995	0.954	0.800	0.600	0.436	0.312	0.141
$1/\sqrt{1-\beta^2} =$	1.005˙	1.048	1.250	1.667	2.294	3.205	7.092

These factors are plotted as a function of β in Fig. 2-4. ◀

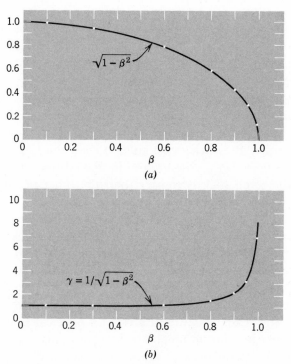

Figure 2-4. (a) A plot of $\sqrt{1-\beta^2}$ as a function of β. (b) A plot of $\gamma = 1/\sqrt{1-\beta^2}$ as a function of β.

2.4 A More Physical Look at the Main Features of the Lorentz Transformation Equations

The main distinguishing features of the Lorentz transformation equations are these: (A) Lengths perpendicular to the relative motion are measured to be the same in both frames; (B) the time interval indicated on a clock is measured to be longer by an observer for whom the clock is moving than by one at rest with respect to the clock; (C) lengths parallel to the relative motion are measured to be contracted compared to the rest lengths by the observer for whom the measured bodies are moving; and (D) two clocks, which are synchronized and separated in one inertial frame, are observed to be out of synchronism from another inertial frame. Here we rederive these features one at a time by thought experiments which focus on the measuring process.

(A) *Comparison of Lengths Perpendicular to the Relative Motion.* Imagine two frames whose relative motion is v along a common x-x' axis. In each frame an observer has a stick extending up from the origin along his vertical (y and y') axis, which he measures to have a (rest) length of exactly one meter, say. As these observers approach and pass each other, we wish to determine whether or not, when the origins coincide, the top ends of the sticks coincide. We can arrange to have the sticks mark each other permanently by a thin pointer at the very top of each (e.g., a razor blade or a paintbrush bristle) as they pass one another. (We displace the sticks very slightly so that they will not collide, always keeping them parallel to the vertical axis.) Notice that the situation is perfectly symmetrical. Each observer claims that his stick is a meter long, each sees the other approach with the same speed v, and each claims that his stick is perpendicular to the relative motion. Furthermore, the two observers must agree on the result of the measurements because they agree upon the simultaneity of the measurements (the measurements occur at the instant the origins coincide). After the sticks have passed, either each observer will find his pointer marked by the other's pointer, or else one observer will find a mark below his pointer, the other observer finding no mark. That is, either the sticks are found to have the same length by both observers, or else there is an absolute result, agreed upon by both observers, that one and the same stick is shorter than the other. That each observer finds the other stick to be the *same* length as his follows at once from the contradiction any other result would indicate with the relativity principle. Suppose, for example, that observer S finds that the S'-stick has left a mark (below his pointer) on his stick. He concludes that the *S'-stick is shorter* than his. This is an absolute result, for the S' observer will find no mark on his stick and will conclude *also* that his stick is shorter. If, instead, the mark was left on the S'-stick, then *each* observer would conclude that the *S-stick* is the *shorter* one. In either case, this would give us a physical basis for preferring one frame over another, for although all the conditions are symmetrical, the results would be unsymmetrical—a result that contradicts the principle of relativity. That is, the laws of physics would not be the same in each inertial frame. We would have a property for detecting absolute motion, in this case; a shrinking stick would mean absolute motion in one direction and a stretching stick would mean absolute motion in the other direction. Hence, to conform to the relativity postulate, we conclude that the length of a body (or space interval) transverse to the relative motion is measured to be the same by all inertial observers.

(B) *Comparison of Time-Interval Measurements.* A simple thought experiment which reveals in a direct way the quantitative relation connecting the time interval between two events as measured from two different inertial frames is the following. Imagine a passenger sitting on a train that moves with uniform velocity v with respect to the ground. The experiment will consist of turning on a flashlight aimed at a mirror directly above on the ceiling and measuring the time it takes the light to travel up and be reflected back down to its starting point. The situation is illustrated in Fig. 2-5. The passenger, who has a wrist watch, say, sees the light ray follow a strictly vertical path (Fig. 2-5*a*) from A to B to C and times the event by his clock (watch). This is a proper time interval, measured by a single clock at one place, the departure and arrival of the light ray occurring at the same place in the

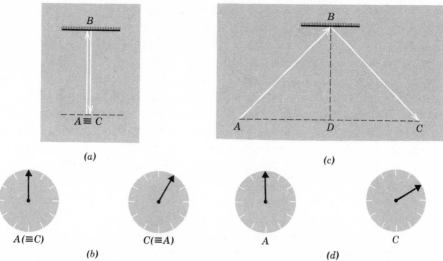

Figure 2-5. (*a*) The path of a light ray as seen by a passenger in the S' frame. B is a mirror on the ceiling, A and C are the *same* point, namely, the bulb of the flashlight, in this frame. (*b*) The readings on the passenger's clock at the start and end of the event, showing the time interval on *one moving clock* (S' frame). (*c*) The path of a light ray as seen by a ground observer (*S*-frame). A and C are the *different locations* of the flashlight bulb at the start and the end of the event, as the train moves to the right with speed v, in this frame. (*d*) Readings on the *two stationary* (synchronized) *clocks* located at the start (A) of the event and the end (C) of the event (S-frame).

passenger's (S') frame. Another observer, fixed to the ground (S) frame, sees the train and passenger move to the right during this interval. He will measure the time interval from the readings on two stationary clocks, one at the position the experiment began (turning-on of flashlight) and a second at the position the experiment ended (arrival of light to flashlight). Hence, he compares the reading of one moving clock (the passenger's watch) to the readings on two stationary clocks. For the S-observer, the light ray follows the oblique path shown in Fig. 2-5c. Thus, the observer on the ground measures the light to travel a greater distance than does the passenger (we have already seen that the transverse distance is the same for each observer). Because the speed of light is the same in both frames, the ground observer sees more time elapse between the departure and the return of the ray of light than does the passenger. He therefore concludes that the passenger's clock runs slow (see Fig. 2-5b and 2-5d). The quantitative result follows at once from the Pythagorean theorem, for

$$\Delta t' = \frac{2BD}{c} \qquad \Delta t = \frac{AB + BC}{c};$$

but

$$(BD)^2 = (AB)^2 - (AD)^2 = (BC)^2 - (DC)^2$$

so that

$$\frac{\Delta t'}{\Delta t} = \frac{2BD}{AB + BC} = \frac{2\sqrt{(AB)^2 - (AD)^2}}{2AB}$$

$$= \sqrt{1 - \left(\frac{AD}{AB}\right)^2} = \sqrt{1 - \frac{v^2}{c^2}}. \qquad (2\text{-}13)$$

Here AD is the horizontal distance travelled at speed v during the time the light travelled with speed c along the hypotenuse. This result is identical to Eqs. 2-11 and 2-12, derived earlier in a more formal way.

(C) *Comparison of Lengths Parallel to the Relative Motion.* The simplest deduction of the length contraction uses the time dilation result just obtained and shows directly that length contraction is a necessary consequence of time dilation. Imagine, for example, that two different inertial observers, one sitting on a train moving through a station with uniform velocity v and the other at rest in the station, want to measure the length of the station's platform. The ground observer measures the length to be L and claims that the passenger covered this distance in a time L/v. This time, Δt, is a nonproper time, for the events observed (passenger passes back end of platform, passenger passes front end of platform) occur at two different places in the ground (S) frame and are timed by two different clocks. The passenger, however, observes the platform approach and recede and finds the two events to occur at the same place in his (S') frame. That is, his clock (wrist watch, say) is located at each event as it occurs. He measures a proper-time interval $\Delta t'$, which, as we have just seen (Eq. 2-13), is related to Δt by $\Delta t' = \Delta t \sqrt{1 - v^2/c^2}$. But $\Delta t = L/v$ so that $\Delta t' = L\sqrt{1 - v^2/c^2}/v$. The passenger claims that the platform moves with the same speed v relative to him so that he would measure the distance from back to front of the platform as $v\,\Delta t'$. Hence, the length of the platform to him is $L' = v\,\Delta t' = L\sqrt{1 - v^2/c^2}$. This is the length-contraction result, namely, that a body of rest length L is measured to have a length $L\sqrt{1 - v^2/c^2}$ parallel to the relative motion in a frame in which the body moves with speed v.

(D) *The Phase Difference in the Synchronization of Clocks.* The student will recall that the Lorentz transformation equation for the time (see Eqs. 2-7 and 2-8) can be written as

$$t = \frac{t' + (v/c^2)x'}{\sqrt{1 - v^2/c^2}}.$$

Here we wish to give a physical interpretation of the vx'/c^2 term, which we call the *phase difference.* We shall synchronize two clocks in one frame and examine what an observer in another frame concludes about the process.

Imagine that we have two clocks, A and B, at rest in the S'-frame. Their separation is L' in this frame. We set off a flashbulb, which is at the exact midpoint, and instruct observers at the clocks to set them to read $t' = 0$ when the light reaches them (see Fig. 2-6a). This is an agreed-upon procedure for synchronizing two separated clocks (see Section 2-1). We now look at this synchronization process as seen by an observer in the S-frame, for whom the clocks A and B move to the right (see Fig. 2-6b) with speed v.

To the S-observer, the separation of the clocks will be $L'\sqrt{1 - v^2/c^2}$. He observes the following sequence of events. The flash goes off and leaves the midpoint traveling in all directions with a speed c. As the wavefront expands at the rate c, the clocks move to the right at the rate v. Clock A intercepts the flash first, before B, and the A observer sets his clock at $t' = 0$ (third picture in sequence). Hence, as far as the S-observer is concerned, A sets his clock to zero time *before* B does and the setting of the primed clocks does

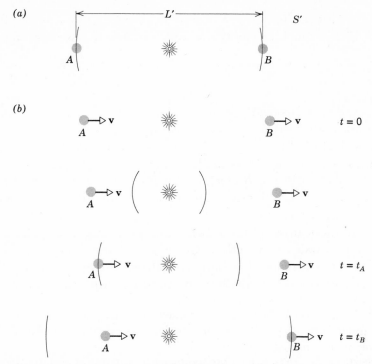

Figure 2-6. (*a*) A flash sent from the midpoint of clocks A and B, at rest in the S'-frame a distance L' apart, arrives simultaneously at A and B. (*b*) The sequence of events as seen from the S-frame, in which the clocks are a distance L apart and move to the right with speed v.

not appear simultaneous to the unprimed observer. Here again we see the relativity of simultaneity; that is, the clocks in the primed frame are *not* synchronized according to the unprimed observer, who uses exactly the same procedure to synchronize his clocks.

By how much do the S'-clocks differ in their readings according to the S-observer? Let $t = 0$ be the time S sees the flash go off. Then, when the light pulse meets clock A, at $t = t_A$, we have

$$ct_A = (L'/2)\sqrt{1 - v^2/c^2} - vt_A.$$

That is, the distance the pulse travels to meet A is less than their initial separation by the distance A travels to the right during this time. When the light pulse later meets clock B (fourth picture in sequence), at $t = t_B$, we have $ct_B = (L'/2)\sqrt{1 - v^2/c^2} + vt_B$. The distance the pulse travels to meet B is greater than their initial separation by the distance B travels to the right during this time. As measured by the clocks in S, therefore, the time interval between the setting of the primed clocks is

$$\Delta t = t_B - t_A = \frac{L'\sqrt{1 - v^2/c^2}/2}{c - v} - \frac{L'\sqrt{1 - v^2/c^2}/2}{c + v}$$

or

$$\Delta t = \frac{L'v\sqrt{1 - v^2/c^2}}{c^2 - v^2}.$$

During this interval, however, S observes clock A to run slow by the factor $\sqrt{1 - v^2/c^2}$ (for "moving clocks run slow") so that to observer S it will read

$$\Delta t' = \Delta t \sqrt{1 - v^2/c^2} = \frac{L'v(1 - v^2/c^2)}{c^2 - v^2} = \frac{L'v}{c^2}$$

when clock B is set to read $t' = 0$.

The result is that the S-observer finds the S' clocks to be out of synchronization, with clock A reading *ahead* in time by an amount $L'v/c^2$. The greater the separation L' of the clocks in the primed frame, the further behind in time is the reading of the B clock as observed at a given instant from the unprimed frame. This is in exact agreement with the Lorentz transformation equation for the time.

Hence, all the features of the Lorentz transformation equations, which we derived in a formal way directly from the postulates of relativity in Section 2-2, can be derived more physically from the measurement processes which were, of course, chosen originally to be consistent with those postulates.

▶ **Example 2.** Why is the fact that simultaneity is not an absolute concept an unexpected result to the classical mind? It is because the speed of light has such a large value compared to ordinary speeds.

Consider these two cases, which are symmetrical in terms of an interchange of the space and time coordinates. *Case 1: S'* observes that two events occur at the same place but are separated in time; S will then declare that the two events occur in different places. *Case 2: S'* observes that two events occur at the same time but are separated in space; S will then declare that the two events occur at different times.

Case 1 is readily acceptable on the basis of daily experience. If a man (S') on a moving train lights two cigarettes, one ten minutes after the other, then these events occur at the same place on *his* reference frame (the train). A ground observer (S), however, would assert that these same events occur at different places in *his* reference system (the ground). Case 2, although true, cannot be easily supported on the basis of daily experience. Suppose that S', seated at the center of a moving railroad car, observes that two men, one at each end of the car, light cigarettes simultaneously. The ground observer S, watching the railroad car go by, would assert (if he could make precise enough measurements) that the man in the back of the car lit his cigarette a little before the man in the front of the car lit his. The fact that the speed of light is so high compared to the speeds of familiar large objects makes Case 2 less intuitively reasonable than Case 1, as we now show.

(a) In Case 1, assume that the time separation in S' is 10 minutes; what is the distance separation observed by S? (b) In Case 2, assume that the distance separation in S' is 25 meters; what is the time separation observed by S? Take $v = 20.0$ m/sec which corresponds to 45 mi/hr or $\beta = v/c = 6.6 \times 10^{-8}$.

(a) From Eqs. 2-8 we have

$$x_2 - x_1 = \frac{x_2' - x_1'}{\sqrt{1 - \beta^2}} + \frac{v(t_2' - t_1')}{\sqrt{1 - \beta^2}}.$$

We are given that $x_2' = x_1'$ and $t_2' - t_1' = 10$ minutes, so that

$$x_2 - x_1 = \frac{(20.0 \text{ m/sec})(10 \text{ min})}{\sqrt{1 - (6.6 \times 10^{-8})^2}} = 12000 \text{ m} = 12 \text{ km}.$$

This result is readily accepted. Because the denominator above is unity for all practical purposes, the result is even numerically what we would expect from the Galilean equations.

(b) From Eqs. 2-8 we have

$$t_2 - t_1 = \frac{t'_2 - t'_1}{\sqrt{1 - \beta^2}} + \frac{(v/c^2)(x'_2 - x'_1)}{\sqrt{1 - \beta^2}}.$$

We are given that $t'_2 = t'_1$ and that $x'_2 - x'_1 = 25$ m, so that

$$t_2 - t_1 = \frac{[(20 \text{ m/sec})/(3.0 \times 10^8 \text{ m/sec})^2](25 \text{ m})}{\sqrt{1 - (6.6 \times 10^{-8})^2}} = 5.6 \times 10^{-15} \text{ sec.}$$

The result is *not* zero, a value that would have been expected by classical physics, but the time interval is so short that it would be very hard to show experimentally that it really was not zero.

If we compare the expressions for $x_2 - x_1$ and for $t_2 - t_1$ above, we see that, whereas v appears as a factor in the second term of the former, v/c^2 appears in the latter. Thus the relatively high value of c puts Case 1 within the bounds of familiar experience but puts Case 2 out of these bounds.

In the following example we consider the realm wherein relativistic effects are easily observable.

Example 3. Among the particles of high-energy physics are charged pions, particles of mass between that of the electron and the proton and of positive or negative electronic charge. They can be produced by bombarding a suitable target in an accelerator with high-energy protons, the pions leaving the target with speeds close to that of light. It is found that the pions are radioactive and, when they are brought to rest, their half-life is measured to be 1.77×10^{-8} secs. That is, half of the number present at any time have decayed 1.77×10^{-8} sec later. A collimated pion beam, leaving the accelerator target at a speed of $0.99c$, is found to drop to half its original intensity 39 m from the target.

(a) Are these results consistent?

If we take the half-life to be 1.77×10^{-8} sec and the speed to be 2.97×10^8 m/sec ($=0.99c$), the distance traveled over which half the pions in the beam should decay is

$$d = vt = 2.97 \times 10^8 \text{ m/sec} \times 1.77 \times 10^{-8} \text{ sec} = 5.3 \text{ m}$$

This appears to contradict the direct measurement of 39 m.

(b) Show how the time dilation accounts for the measurements.

If the relativistic effects did not exist, then the half-life would be measured to be the same for pions at rest and pions in motion (an assumption we made in part *a* above). In relativity, however, the nonproper and proper half-lives are related by

$$\Delta t = \frac{\Delta \tau}{\sqrt{1 - v^2/c^2}}.$$

The proper time in this case is 1.77×10^{-8} sec, the time interval measured by a clock attached to the pion, that is, at one place in the rest frame of the pion. In the laboratory frame, however, the pions are moving at high speeds and the time interval there (a nonproper one) will be measured to be larger (moving clocks appear to run slow). The nonproper half-life, measured by two different clocks in the laboratory frame, would then be

$$\Delta t = \frac{1.77 \times 10^{-8} \text{ sec}}{\sqrt{1 - (0.99)^2}} = 1.3 \times 10^{-7} \text{ sec.}$$

This is the half-life appropriate to the laboratory reference frame. Pions that live this long, traveling at a speed $0.99c$, would cover a distance

$$d = 0.99c \times \Delta t = 2.97 \times 10^{-8} \text{ m/sec} \times 1.3 \times 10^{-7} \text{ sec} = 39 \text{ m,}$$

exactly as measured in the laboratory.

(c) Show how the length contraction accounts for the measurements.

In part a we used a length measurement (39 m) appropriate to the laboratory frame and a time measurement (1.77×10^{-8} sec) appropriate to the pion frame and incorrectly combined them. In part b we used the length (39 m) and time (1.3×10^{-7} sec) measurements appropriate to the laboratory frame. Here we use length and time measurements appropriate to the pion frame.

We already know the half-life in the pion frame, that is, the proper time 1.77×10^{-8} sec. What is the distance covered by the pion beam during which its intensity falls to half its original value? If we were sitting on the pion, the laboratory distance of 39 m would appear much shorter to us because the laboratory moves at a speed $0.99c$ relative to us (the pion). In fact, we would measure the distance

$$d' = d\sqrt{1 - v^2/c^2} = 39\sqrt{1 - (0.99)^2} \text{ m.}$$

The time elapsed in covering this distance is $d'/0.99c$ or

$$\Delta \tau = \frac{39 \text{ m} \sqrt{1 - (0.99)^2}}{0.99c} = 1.77 \times 10^{-8} \text{ sec,}$$

exactly the measured half-life in the pion frame.

Thus, depending on which frame we choose to make measurements in, this example illustrates the physical reality of either the time-dilation or the length-contraction predictions of relativity. Each pion carries its own clock, which determines the proper time τ of decay, but the decay time observed by a laboratory observer is much greater. Or, expressed equivalently, the moving pion sees the laboratory distances contracted and in its proper decay time can cover laboratory distances greater than those measured in its own frame.

Notice that in this region of $v \approx c$ the relativistic effects are large. There can be no doubt whether in our example, the distance is 39 m or 5.3 m. If the proper time were applicable to the laboratory frame, the time (1.3×10^{-7} sec) to travel 39 m would correspond to over seven half-lives (i.e., 1.3×10^{-7} sec$/1.8 \times 10^{-8}$ sec $\cong 7$). Instead of the beam being reduced to half its original intensity, it would be reduced to $(1/2)^7$ or $1/128$ its original intensity in travelling 39 m. Such differences are very easily detectable.

This example is by no means an isolated result (see, e.g., Problems 27 to 30 and Ref. 4). All the kinematic (and dynamic) measurements in high-energy physics are consistent with the time-dilation and length-contraction results. The experiments and the accelerators themselves are designed to take relativistic effects into account. Indeed, relativity is a routine part of the everyday world of high-speed physics and engineering. ◀

2.5 The Observer in Relativity

There are many shorthand expressions in relativity which can easily be misunderstood by the uninitiated. Thus the phrase "moving clocks run slow" means that a clock moving at a constant velocity relative to an inertial frame containing synchronized clocks will be found to run slow *when timed by those clocks*. We compare *one moving clock* with *two synchronized stationary clocks*. Those who assume that the phrase means anything else often encounter difficulties.

Similarly, we often refer to "an observer." The meaning of this term also is quite definite, but it can be misinterpreted. *An observer is really an infinite set of recording clocks distributed throughout space, at rest and synchronized with respect to one another.* The space-time coordinates of an event (x,y,z,t) are recorded by the clock at the location (x,y,z) of the event at the time (t) it occurs. Measurements thus recorded throughout space-time (we might call them local measurements) are then available to be picked up and analyzed by an experimenter. Thus, the observer can also be thought of as the experimenter who collects the measurements made in this way. Each inertial frame is imagined to have such a set of recording clocks, or such an observer. The relations between the space-time coordinates of a physical event measured by one observer (S) and the space-time coordinates of the *same* physical event measured by another observer (S') are the equations of transformation.

A misconception of the term "observer" arises from confusing "measuring" with "seeing." For example, it had been commonly assumed for some time that the relativistic length contraction would cause rapidly moving objects to appear to the eye to be shortened in the direction of motion. The location of all points of the object measured at the same time would give the "true" picture according to our use of the term "observer" in relativity. But, in the words of V. F. Weisskopf [5]:

"When we see or photograph an object, we record light quanta emitted by the object when they arrive simultaneously at the retina or at the photographic film. This implies that these light quanta have *not* been emitted simultaneously by all points of the object. The points further away from the observer have emitted their part of the picture earlier than the closer points. Hence, if the object is in motion, the eye or the photograph gets a distorted picture of the object, since the object has been at different locations when different parts of it have emitted the light seen in the picture."

To make a comparison with the relativistic predictions, therefore, we must first allow for the time of flight of the light quanta from the different parts of the object. Without this correction, we see a distortion due to *both* the optical *and* the relativistic effects. Circumstances sometimes exist in which the object appears to have suffered no contraction at all. Under other special circumstances the Lorentz contraction can be seen unambiguously [see Refs. 6 and 7]. But the term "observer" does *not* mean "viewer" in relativity and we shall continue to use it only in the sense of "measurer" described above.

2.6 The Relativistic Addition of Velocities

In classical physics, if we have a train moving with a velocity **v** with respect to ground and a passenger on the train moves with a velocity **u′** with respect to the train, then the passenger's velocity relative to the ground **u** is just the vector sum of the two velocities (see Eq. 1-5), that is,

$$\mathbf{u} = \mathbf{u'} + \mathbf{v}. \tag{2-15}$$

This is simply the classical, or Galilean, velocity addition theorem (see *Physics,* Part I, Sec. 4-6). How do velocities add in special relativity theory?

Consider, for the moment, the special case wherein all velocities are along the common x-x′ direction of two inertial frames S and S′. Let S be the ground frame and S′ the frame of the train, whose speed relative to the ground is v (see Fig. 2-7). The passenger's speed in the S′-frame is u′, and his position on the train as time goes on can be described by $x' = u't'$. What is the speed of the passenger observed from the ground? Using the Lorentz transformation equations (Eqs. 2-7), we have

$$x' = \frac{x - vt}{\sqrt{1 - v^2/c^2}} = u't' \quad \text{and} \quad t' = \frac{t - (v/c^2)x}{\sqrt{1 - v^2/c^2}}.$$

Combining these yields

$$x - vt = u'\left(t - \frac{v}{c^2}x\right),$$

which can be written as

$$x = \frac{(u' + v)}{(1 + u'v/c^2)}t. \tag{2-16}$$

If we call the passenger's speed relative to ground u, then his ground location as time goes on is given by $x = ut$. Comparing this to Eq. 2-16, we obtain

$$u = \frac{u' + v}{1 + u'v/c^2} \tag{2-17}$$

This is the *relativistic,* or Einstein *velocity addition theorem.*

If u′ and v are very small compared to c, Eq. 2-17 reduces to the classical result, Eq. 2-15, $u = u' + v$, for then the second term in the denominator of Eq. 2-17 is negligible compared to one. On the other hand, if $u' = c$, it always follows that $u = c$ no matter what the value of v. Of course, $u' = c$ means that our "passenger" is a light pulse and we know that an assumption used to derive the transformation formulas was exactly this result; that is,

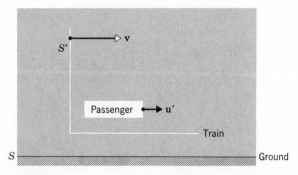

Figure 2-7. A schematic view of the system used in deriving the equations for the relativistic addition of velocities.

that all observers measure the same speed c for light. Formally, we get, with $u' = c$,

$$u = \frac{c + v}{1 + cv/c^2} = \frac{c + v}{c(c + v)}c^2 = c.$$

Hence, any velocity (less than c) relativistically added to c gives a resultant c. In this sense, c plays the same role in relativity that an infinite velocity plays in the classical case.

The Einstein velocity addition theorem can be used to explain the observed result of the experiments designed to test the various emission theories of Chapter One. The basic result of these experiments is that the velocity of light is independent of the velocity of the source (see Section 1-8). We have seen that this is a basic postulate of relativity so that we are not surprised that relativity yields agreement with these experiments. If, however, we merely looked at the formulas of relativity, unaware of their physical origin, we could obtain this specific result from the velocity addition theorem directly. For, let the source be the S' frame. In that frame the pulse (or wave) of light has a speed c in vacuum according to the emission theories. Then, the pulse (or wave) speed measured by the S-observer, for whom the source moves, is given by Eq. 2-17, and is also c. That is, $u = c$ when $u' = c$, as shown above.

It follows also from Eq. 2-17 that the addition of two velocities, each smaller than c, cannot exceed the velocity of light.

◆ **Example 4.** In Example 2 of Chapter One, we found that when two electrons leave a radioactive sample in opposite directions, each having a speed $0.67c$ with respect to the sample, the speed of one electron relative to the other is $1.34c$ according to classical physics. What is the relativistic result?

We may regard one electron as the S-frame, the sample as the S'-frame, and the other electron as the object whose speed in the S-frame we seek (see Fig. 1-3). Then

$$u' = 0.67c \qquad v = 0.67c$$

and
$$u = \frac{u' + v}{1 + u'v/c^2} = \frac{(0.67 + 0.67)c}{1 + (0.67)^2} = \frac{1.34}{1.45}c = 0.92c.$$

The speed of one electron relative to the other is less than c.

Does the relativistic velocity addition theorem alter the result of Example 1 of Chapter One? Explain.

Example 5. Show that the Einstein velocity addition theorem leads to the observed Fresnel drag coefficient of Eq. 1-11.

In this case, v_w is the velocity of water with respect to the apparatus and c/n is the velocity of light relative to the water. That is, in our formula we have

$$u' = \frac{c}{n} \qquad \text{and} \qquad v = v_w.$$

Then, the velocity of light relative to the apparatus is

$$u = \frac{c/n + v_w}{1 + v_w/nc}$$

For v_w/c small (in the experiments $v_w/c = 2.3 \times 10^{-8}$) we can neglect terms of second order in v_w/c, so that, using the binomial expansion, we have

$$u \cong \left(\frac{c}{n} + v_w\right)\left(1 - \frac{v_w}{nc}\right) \cong \frac{c}{n} + v_w\left(1 - \frac{1}{n^2}\right).$$

This is exactly Eq. 1-11, the observed first-order effect. Notice that there is no need to assume any "drag" mechanism, or to invent theories on the interaction between matter and the "ether." The result is an inevitable consequence of the velocity addition theorem and illustrates the powerful simplicity of relativity. ◀

It is interesting and instructive to note that there *are* speeds in excess of c. Although matter or energy (i.e., signals) cannot have speeds greater than c, certain kinematical processes *can* have super-light speeds (see Ref. 8 and Question 20). For example, the succession of points of intersection of the blades of a giant scissors, as the scissors is rapidly closed, may be generated at a speed greater than c. Here geometrical points are involved, the motion being an illusion, whereas the material objects involved (atoms in the scissors blades, e.g.) always move at speeds less than c. Other similar examples are the succession of points on a fluorescent screen as an electron beam sweeps across the screen, or the light of a searchlight beam sweeping across the cloud cover in the sky. The electrons, or the light photons, which carry the energy, move at speeds not exceeding c.

Thus far, we have considered only the transformation of velocities parallel to the direction of relative motion of the two frames of reference (the x-x' direction). To signify this, we should put x subscripts on u and u' in Eq. 2-17, obtaining

$$u_x = \frac{u'_x + v}{1 + u'_x(v/c^2)} \tag{2-18}$$

For velocities that are perpendicular to the direction of relative motion, the result is more involved. Imagine that an object moves parallel to the y'-axis in S'. Let it be observed to be at y'_1 and y'_2 at the times t'_1 and t'_2, respectively, so that its velocity in S' is $u'_y = \Delta y'/\Delta t' = (y'_2 - y'_1)/(t'_2 - t'_1)$. To find its velocity in S, we use the Lorentz transformation equations and obtain

$$y'_2 - y'_1 = y_2 - y_1$$

$$t'_2 - t'_1 = \frac{t_2 - t_1 - (x_2 - x_1)v/c^2}{\sqrt{1 - v^2/c^2}} = \frac{\Delta t - \Delta x(v/c^2)}{\sqrt{1 - v^2/c^2}}$$

so that

$$\frac{\Delta y'}{\Delta t'} = \frac{\Delta y \sqrt{1 - v^2/c^2}}{\Delta t - \Delta x(v/c^2)} = \frac{(\Delta y/\Delta t)\sqrt{1 - v^2/c^2}}{1 - \left(\frac{\Delta x}{\Delta t}\right)v/c^2}.$$

Now $\Delta y/\Delta t$ is u_y and $\Delta x/\Delta t$ is u_x so that

$$u'_y = \frac{u_y \sqrt{1 - v^2/c^2}}{1 - u_x(v/c^2)}.$$

For comparison with Eq. 2-18, we can write the corresponding inverse transformation. We merely change v to $-v$ and interchange primed and unprimed quantities, obtaining

$$u_y = \frac{u'_y \sqrt{1 - v^2/c^2}}{1 + u'_x(v/c^2)}. \tag{2-19}$$

The student can derive the result also by seeking $\Delta y/\Delta t$ directly, instead of $\Delta y'/\Delta t'$ as is done above (see Problem 31). In exactly the same way, we also find

$$u_z = \frac{u'_z \sqrt{1 - v^2/c^2}}{1 + u'_x(v/c^2)}. \tag{2-20}$$

In Table 2-2 we summarize the relativistic velocity transformation equations. We shall have occasion to use these results, and to interpret them further, in later sections. For the moment, however, let us note certain aspects of the transverse velocity transformations. The perpendicular, or transverse, components (i.e., u_y and u_z) of the velocity of an object as seen in the S-frame are related both to the transverse components (i.e., u'_y and u'_z) and to the parallel component (i.e., u'_x) of the velocity of the object in the S'-frame. The result is not simple because neither observer is a proper one. If we choose a frame in which $u'_x = 0$, however, then the transverse results become $u_z = u'_z \sqrt{1 - v^2/c^2}$ and $u_y = u'_y \sqrt{1 - v^2/c^2}$. But no length contraction is involved for transverse space intervals, so what is the origin of the $\sqrt{1 - v^2/c^2}$ factor? We need only point out that velocity, being a ratio of length interval to time interval, involves the time coordinate too, so that time dilation is involved. Indeed, this special case of the transverse velocity transformation is a direct time-dilation effect.

2.7 Aberration and Doppler Effect in Relativity

Up to now we have shown how relativity can account for the experimental results of various light-propagation experiments listed in Table 1-2 (e.g., the Fresnel drag coefficient and the Michelson-Morley result) and at the same time how it predicts new results also confirmed by experiment (time dilation in pion or meson decay, also in Table 1-2). Here we deduce the aberration result. In doing this, we shall also come upon another new result predicted

Table 2-2 THE RELATIVISTIC VELOCITY TRANSFORMATION EQUATIONS

$u'_x = \dfrac{u_x - v}{1 - u_x v/c^2}$	$u_x = \dfrac{u'_x + v}{1 + u'_x v/c^2}$
$u'_y = \dfrac{u_y \sqrt{1 - v^2/c^2}}{1 - u_x v/c^2}$	$u_y = \dfrac{u'_y \sqrt{1 - v^2/c^2}}{1 + u'_x v/c^2}$
$u'_z = \dfrac{u_z \sqrt{1 - v^2/c^2}}{1 - u_x v/c^2}$	$u_z = \dfrac{u'_z \sqrt{1 - v^2/c^2}}{1 + u'_x v/c^2}$

by relativity and confirmed by experiment, namely a transverse Doppler effect.

Consider a train of plane monochromatic light waves of unit amplitude emitted from a source at the origin of the S'-frame, as shown in Fig. 2-8. The rays, or wave normals, are chosen to be in (or parallel to) the x'-y' plane, making an angle θ' with the x'-axis. An expression describing the propagation would be of the form

$$\cos 2\pi \left[\frac{x' \cos \theta' + y' \sin \theta'}{\lambda'} - \nu't' \right], \qquad (2\text{-}21)$$

for this is a single periodic function, amplitude unity, representing a wave moving with velocity $\lambda'\nu' \ (= c)$ in the θ'-direction. Notice, for example, that for $\theta' = 0$ it reduces to $\cos 2\pi[x'/\lambda' - \nu't']$ and for $\theta' = \pi/2$ it reduces to $\cos 2\pi[y'/\lambda' - \nu't']$, well-known expressions for propagation along the positive-x' and positive-y' directions, respectively, of waves of frequency ν' and wavelength λ'. The alternate forms, $\cos (2\pi/\lambda')[x' - \lambda'\nu't']$ and $\cos (2\pi/\lambda') [y' - \lambda'\nu't']$ show that the wave speed is $\lambda'\nu'$ which, for electromagnetic waves, is equal to c.

In the S-frame these wavefronts will still be planes, for the Lorentz transformation is linear so that a plane transforms into a plane. Hence, in the unprimed, or S, frame the expression describing the propagation will have the same form:

$$\cos 2\pi \left[\frac{x \cos \theta + y \sin \theta}{\lambda} - \nu t \right] \qquad (2\text{-}22)$$

Here, λ and ν are the wavelength and frequency, respectively, measured in the S-frame, and θ is the angle a ray makes with the x-axis. We know, if

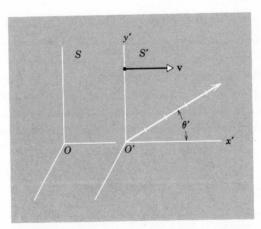

Figure 2-8. A ray, or wave normal, of plane monochromatic light waves is emitted from the origin of the S' frame. The bars signify wavefronts separated by one wavelength from adjacent wavefronts. The direction of propagation makes an angle θ' with the x'-axis, the rays being parallel to the x'-y' plane.

expressions 2-21 and 2-22 are to represent electromagnetic waves, that $\lambda\nu = c$, just as $\lambda'\nu' = c$, for c is the velocity of electromagnetic waves, the same for each observer.

Now let us apply the Lorentz transformation equations directly to expression 2-21, putting

$$x' = \frac{x - vt}{\sqrt{1 - \beta^2}}, \qquad y' = y, \qquad \text{and} \qquad t' = \frac{t - (v/c^2)x}{\sqrt{1 - \beta^2}}.$$

We obtain

$$\cos 2\pi \left[\frac{1}{\lambda'} \frac{(x - vt)}{\sqrt{1 - \beta^2}} \cos\theta' + \frac{y \sin\theta'}{\lambda'} - \nu' \frac{[t - (v/c^2)x]}{\sqrt{1 - \beta^2}} \right]$$

or, on rearranging terms and using $\lambda'\nu' = c$,

$$\cos 2\pi \left[\frac{\cos\theta' + \beta}{\lambda'\sqrt{1 - \beta^2}} x + \frac{\sin\theta'}{\lambda'} y - \frac{(\beta \cos\theta' + 1)\nu'}{\sqrt{1 - \beta^2}} t \right]$$

As expected, this has the form of a plane wave in the *S*-frame and must be identical to expression 2-22, which represents the same thing. Hence, the coefficient of x, y, and t in each expression must be equated, giving us

$$\frac{\cos\theta}{\lambda} = \frac{\cos\theta' + \beta}{\lambda'\sqrt{1 - \beta^2}} \tag{2-23}$$

$$\frac{\sin\theta}{\lambda} = \frac{\sin\theta'}{\lambda'} \tag{2-24}$$

$$\nu = \frac{\nu'(1 + \beta \cos\theta')}{\sqrt{1 - \beta^2}} \tag{2-25}$$

We also have the relation

$$\lambda\nu = \lambda'\nu' = c, \tag{2-26}$$

a condition we knew in advance.

In the procedure that we have adopted here, we start with a light wave in S' for which we know λ', ν', and θ' and we wish to find what the corresponding quantities λ, ν, and θ are in the *S*-frame. That is, we have three unknowns but we have four equations (Eqs. 2-23 to 2-26) from which to determine the unknowns. The unknowns have been overdetermined, which means simply that the equations are not all independent. If we eliminate one equation, for instance, by dividing one by another (i.e., we combine two equations), we shall obtain three independent relations. It is simplest to divide Eq. 2-24 by Eq. 2-23; this gives us

$$\tan\theta = \frac{\sin\theta' \sqrt{1 - \beta^2}}{\cos\theta' + \beta} \tag{2-27a}$$

which is *the relativistic equation for the aberration of light*. It relates the directions of propagation, θ and θ', as seen from two different inertial frames. The inverse transformation can be written at once as

$$\tan \theta' = \frac{\sin \theta \sqrt{1 - \beta^2}}{\cos \theta - \beta} \tag{2-27b}$$

wherein β of Eq. 2-27a becomes $-\beta$ and we interchange primed and un-primed quantities. Experiments in high-energy physics involving photon emission from high velocity particles confirm the relativistic formula exactly.

▶ **Example 6.** Show that the observed first-order aberration effect, which corresponds to the classical picture, is a special case of the exact relativistic formula.

Consider the case of a star directly overhead in the S-frame. One receives plane waves whose direction of propagation is along the negative y direction. Hence, $\theta = 3\pi/2$. In S', the propagation direction is θ', given by Eq. 2-27b with $\theta = 3\pi/2$. That is,

$$\tan \theta' = \frac{\sin (3\pi/2) \sqrt{1 - \beta^2}}{\cos (3\pi/2) - \beta} = \frac{-\sqrt{1 - \beta^2}}{-\beta}.$$

When v is very small compared to c ($v \ll c$), then v/c, or β, is very small compared to one. Thus, β^2 will be negligible compared to one; neglecting terms in the second order then, we can write

$$\tan \theta' = \frac{-\sqrt{1 - \beta^2}}{-\beta} \cong \frac{-1}{-\beta} = \frac{1}{\beta} = \frac{c}{v}.$$

This result is in perfect agreement with the observed first-order aberration effect, corresponding to the classical interpretation of the situation, as shown in Fig. 2-9. In Fig. 2-9a we show the propagation direction of the starlight in S and in S' and in Fig. 2-9b the orientation of the telescopes in S and S' which observe the star. ◀

The third of our four equations above (Eqs. 2-23 to 2-26) gives us directly the one remaining phenomenon we promised to discuss; that is, *the relativistic equation for the Doppler effect,*

$$\nu = \frac{\nu'(1 + \beta \cos \theta')}{\sqrt{1 - \beta^2}} \tag{2-25a}$$

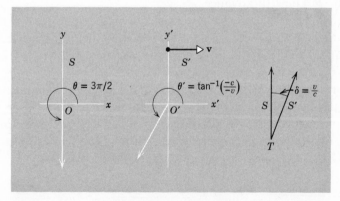

Figure 2-9. (*a*) In S, the direction of propagation from the source is along $-y$, $\theta = 3\pi/2$. In S', the same ray makes an angle θ' with the y'-axis. (*b*) The line of sight of the telescope in S is vertical and in S' is inclined forward by an angle $\delta = v/c$, in order to see the source.

which we can also write inversely as

$$\nu' = \frac{\nu(1 - \beta \cos \theta)}{\sqrt{1 - \beta^2}}. \tag{2-25b}$$

Let us first check that the relativistic formula reduces to the classical one. That is, for $v \ll c$ we can neglect terms higher than first order in v/c, or β, and the first-order result should be the classical one. From Eq. 2-25b, we get (using the binomial theorem expansion twice through first-order terms)

$$\nu = \frac{\nu' \sqrt{1 - \beta^2}}{1 - \beta \cos \theta} \cong \frac{\nu'}{1 - \beta \cos \theta} \cong \nu'(1 + \beta \cos \theta)$$

which is the classical result (see *Physics,* Part I, Sec. 20-7). This becomes clear on consideration of the more familiar special cases. For, with $\theta = 0$, which corresponds to observer S seeing the source move toward him or his moving toward the source, we obtain

$$\nu = \nu'(1 + \beta) = \nu'\left(1 + \frac{v}{c}\right)$$

which shows that the observed frequency ν is greater than the proper frequency ν'. With $\theta = 180°$, which corresponds to observer S seeing the source move away from him or his moving away from the source, we obtain

$$\nu = \nu'(1 - \beta) = \nu'\left(1 - \frac{v}{c}\right)$$

which shows that the observed frequency ν is less than the proper frequency ν'. Finally, for $\theta = 90°$, wherein the line of sight is perpendicular to the relative motion, there is no Doppler effect classically; that is, $\nu = \nu'$. All these first-order results are classical effects.

Now, if v is *not* small compared to c, we should obtain relativistic (second-order) effects. It is convenient to think of these effects separately as a longitudinal one and a transverse one. Thus, for the *longitudinal Doppler effect in relativity,* we use Eq. 2-25b and set $\theta = 0$ or $\theta = 180°$. That is, in $\nu = (\nu' \sqrt{1 - \beta^2})/(1 - \beta \cos \theta)$, with $\theta = 0$ (source and observer move *toward* one another) we obtain

$$\nu = \nu' \sqrt{\frac{1 + \beta}{1 - \beta}} = \nu' \sqrt{\frac{c + v}{c - v}}; \tag{2-28}$$

and with $\theta = 180°$ (source and observer move *away* from one another) we obtain

$$\nu = \nu' \sqrt{\frac{1 - \beta}{1 + \beta}} = \nu' \sqrt{\frac{c - v}{c + v}} \tag{2-29}$$

These results, were first confirmed experimentally in 1938 by Ives and Stilwell, who (following a suggestion first made by Einstein in 1907) used a beam of excited hydrogen atoms of well-defined speed and direction as the source of radiation [9, 10]. The experiment was repeated in 1961 with higher accuracy by Mandelberg and Witten [11], again confirming the relativistic effect.

More striking, however, is the fact that the relativistic formula predicts a *transverse Doppler effect,* an effect that is *purely relativistic,* for there is no transverse Doppler effect in classical physics at all. This prediction follows from Eq. 2-25b, $\nu = (\nu' \sqrt{1 - \beta^2})/(1 - \beta \cos \theta)$, when we set $\theta = 90°$, obtaining

$$\nu = \nu' \sqrt{1 - \beta^2}. \tag{2-30}$$

If our line of sight is 90° to the relative motion, then we should observe a frequency ν which is *lower* than the proper frequency ν' of the source which is sweeping by us. Ives and Stilwell [9] in 1938 and 1941, and Otting [12] in 1939 confirmed the existence of this transverse Doppler effect, and more recently Kundig [13] obtained excellent quantitative data comfirming the relativistic formula to within the experimental error of 1.1 percent.

It is instructive to note that the transverse Doppler effect has a simple time-dilation interpretation. The moving source is really a moving clock, beating out electromagnetic oscillations. We have seen that moving clocks appear to run slow. Hence, we see a given number of oscillations in a time that is longer than the proper time. Or, equivalently, we see a smaller number of oscillations in our unit time than is seen in the unit time of the proper frame. Therefore, we observe a lower frequency than the proper frequency. The transverse Doppler effect is another physical example confirming the relativistic time dilation.

In both the Dopper effect and aberration, the theory of relativity introduces an intrinsic simplification over the classical interpretation of these effects in that the two separate cases which are different in classical theory (namely, source at rest-moving observer and observer at rest-moving source) are identical in relativity. This, too, is in accord with observation. Notice, also, that a single derivation yields at once three effects, namely aberration, longitudinal Doppler effect, and transverse Doppler effect.

2.8   The Common Sense of Special Relativity

We are now at a point where a retrospective view can be helpful. Special relativity theory makes still more predictions than we have discussed so far that contradict classical views. Later we shall see that, in those cases too, experimental results confirm the relativistic predictions. Indeed, throughout atomic, nuclear, high-energy, and solid-state physics, relativity is used in an almost commonplace way as the correct description of the real microscopic world. Furthermore, relativity is a consistent theory, as we have shown already in many ways and shall continue to show later. However, because our everyday macroscopic world is classical to a good approximation and students have not yet lived with or used relativity enough to become sufficiently familiar with it, there may remain misconceptions about the theory which are worth discussing now.

(A) *The Limiting Speed c of Signals.* We have seen that, if it were possible to transmit signals with infinite speed, we could establish in an absolute way whether or not two events are simultaneous. The relativity of simultaneity depended on the existence of a finite speed of transmission of signals. Now we probably would grant that it is unrealistic to expect that any physical action could be transmitted with infinite speed. It does indeed seem fanciful that we could initiate a signal that would travel to all parts of our universe in zero time. It is really the classical physics (which at bottom makes such an assumption) that is fictitious (science fiction) and not the relativistic physics, which postulates a limiting speed. Furthermore, when experiments are carried out, the relativity of time measurements is confirmed. Nature does indeed show that relativity is a practical theory of measurement and not a philosophically idealistic one, as is the classical theory.

We can look at this in another way. From the fact that experiment denies the absolute nature of time, we can conclude that signals cannot be transmitted with infinite speed. Hence, there must be a certain finite speed that cannot be exceeded and which we call the limiting speed. The principle of relativity shows at once that this limiting speed is the speed of light, since the result that no speed can exceed a given limit is certainly a law of physics and, according to the principle of relativity, the laws of physics are the same for all inertial observers. Therefore this given limit, the limiting speed, must be exactly the same in all inertial reference frames. We have seen, from experiment, that the speed of light has exactly this property.

Viewed in this way, the speed of electromagnetic waves in vacuum assumes a role wider than the travel rate of a particular physical entity. It becomes instead a limiting speed for the motion of anything in nature.

(B) *Absolutism and Relativity.* The theory of relativity could have been called, instead, the theory of absolutism with some justification.

The fact that the observers who are in relative motion assign different numbers to length and time intervals between the pair of events, rather than finding these numbers to be absolutes, upsets the classical mind. This is so in spite of the fact that even in classical physics the measured values of the momentum or kinetic energy of a particle, for example, also are different for two observers who are in relative motion. What is troublesome, apparently, is the philosophic notion that length and time in the abstract are absolute quantities and the belief that relativity contradicts this notion. Now, without going into such a philosophic byway, it is important to note that relativity simply says that the *measured* length or time interval between a pair of events is affected by the relative *motion* of the events and measurer. Relativity is a theory of measurement, and motion affects measurement. Let us look at various aspects of this.

That relative motion should affect measurements is almost a "common-sense" idea—classical physics is full of such examples, including the aberration and Doppler effects already discussed. Furthermore, to explain such phenomena in relativity, we need not talk about the structure of matter or the idea of an ether in order to find changes in length and duration due to motion. Instead, the results follow directly *from the measurement process itself.* Indeed, we find that the phenomena are *reciprocal.* That is, just as A's clock seems to B to run slow, so does B's clock seem to run slow to A; just as A's meter stick seems to B to have contracted in the direction of motion, so likewise B's meter stick seems to A to have contracted in exactly the same way.

Moreover, there *are* absolute lengths and times in relativity. The *rest length* of a rod is an absolute quantity, the same for all inertial observers: If a given rod is measured by different inertial observers by bringing the rod to rest in their respective frames, each will measure the same length. Similarly the clocks, the *proper time* (which might better have been called "local time") is an invariant quantity:* the frequency of oscillation of an ammonia molecule, for instance, would be measured to be the same by different inertial observers who bring the molecule to rest in their respective frames.

Where relativity theory is clearly "more absolute" than classical physics is in the relativity principle itself: the *laws of physics* are absolute. We have seen that the Galilean transformations and classical notions contradicted the invariance of electromagnetic (and optical) laws, for example. Surely, giving up the absoluteness of the laws of physics, as classical notions of time and length demand, would leave us with an arbitrary and complex physical world. By comparison, relativity is absolute and simple.

(C) *The "Reality" of the Length Contraction.* Is the length contraction "real" or apparent? We might answer this by posing a similar question. Is the frequency, or wavelength, shift in the Doppler effect real or apparent? Certainly the proper frequency (i.e., the rest frequency) of the source is measured to be the same by all observers who bring the source to rest before taking the measurement. Likewise, the proper length is invariant. When the source and observer are in relative motion, the observer definitely measures a frequency (or wavelength) shift. Likewise, the moving rod is definitely measured to be contracted. The effects are real in the same sense that the measurements are real. We do not claim that the proper frequency has changed because of our measured shift. Nor do we claim that the proper length has changed because of our measured contraction. The effects are apparent (i.e., caused by the motion) in the same sense that proper quantities have not changed.

We do not speak about theories of matter to explain the contraction but, instead, we invoke the measurement process itself. For example, we do not assert, as Lorentz sought to prove, that motion produces a physical contraction through an effect on the elastic forces in the electronic or atomic constitution of matter (motion is *relative*, not absolute), but instead we remember the fish story. If a fish is swimming in water and his length is the distance between his tail and his nose, measured simultaneously, observers who disagree on whether measurements are simultaneous or not will certainly disagree on the measured length. Hence, length contraction is due to the relativity of simultaneity.

Since length measurements involve a comparison of two lengths (moving rod and measuring rod, e.g.) we can see that the Lorentz length contraction is really not a property of a single rod by itself but instead is a relation between two such rods in relative motion. The relation is both observable and reciprocal.

(D) *Rigid Bodies and Unit Length.* In classical physics, the notion of an ideal rigid body was often used as the basis for length (i.e., space) measurements. In principle, a rigid rod of unit length is used to lay out a distance scale. Even in relativity we can imagine a standard rod defining a unit distance, this same rod being brought to rest in each observer's frame to lay out

---

*In terms of simultaneity, we can say that the time order of two events *at the same place* can be absolutely determined. It is in the case that two events are separated in space that simultaneity is a relative concept.

space-coordinate units. However, the concept of an ideal rigid body is unten-able in relativity, for such a body would be capable of transmitting signals instantaneously; a disturbance at one end would be propagated with infinite velocity through the body, in contradiction to the relativistic principle that there is a finite upper limit to the speed of transmission of a signal.

Conceptually, then, we must give up the notion of an ideal rigid body. This causes no problems for, at bottom, time measurements are primary and space measurements secondary. We know that this is so in relativity (the simultaneity concept is used in the definition of length) but it is less well recognized that a similar situation exists in classical physics.

For example, we do not use the rigid-body concept in making distance measurements on the astronomical scale. Instead we use the "radar" method. We measure the round-trip time for electromagnetic waves and derive dis-tance from a product of the velocity c and the time interval. Even the units, such as light-years, suggest this procedure. An analogous "sonar" technique is used by animals (e.g., bats and fish) for distance measurement. And on the atomic and subatomic scale we do not invoke rigid bodies for distance measurements either. We again use the properties of electromagnetic waves and not of rigid bodies. Indeed, the very quantity that is today taken as the unit of length is the wavelength of light of a given frequency ν, the wavelength being the distance c/ν traveled in one period at a speed c. In atomic theory, the frequencies are the standard or characteristic quantities, so that the time standards are primary and lengths are determined from them by the use of c.

It is fitting, in emphasizing the common sense of relativity, to conclude with this quotation from Bondi [14] on the presentation of relativity theory:

"At first, relativity was considered shocking, anti-establishment and highly mysterious, and all presentations intended for the population at large were meant to emphasize these shocking and mysterious aspects, which is hardly conducive to easy teaching and good understanding. They tended to empha-size the revolutionary aspects of the theory whereas, surely, it would be good teaching to emphasize the continuity with earlier thought. . . .

"It is first necessary to bring home to the student very clearly the Newtonian attitude. Newton's first law of dynamics leads directly to the notion of an *inertial observer*, defined as an observer who finds the law of inertia to be correct. . . . The utter equivalence of inertial observers to each other for the purpose of Newton's first law is a direct and logical consequence of this law. The equivalence with regard to the second law is not a logical necessity but a very plausible extension, and with this plausible extension we arrive at Newton's principle of relativity: *that all inertial observers are equivalent as far as dynamical experiments go.* It will be obvious that the restriction to dynamical experiments is due simply to this principle of relativity having been derived from the laws of dynamics. . . .

"The next step . . . is to point out how absurd it would be if dynamics were in any sense separated from the rest of physics. There is no experiment in physics that involves dynamics alone and nothing else. . . . Hence, New-ton's principle of relativity is empty because it refers only to a class of experiment that does not exist—the purely dynamical experiment. The choice is therefore presented of either throwing out this principle or removing its restriction to dynamical experiments. The first alternative does not lead us any further, and clearly disregards something of significance in our experience. The second alternative immediately gives us Einstein's principle of relativity: *that all inertial observers are equivalent*. It presents this principle, not as a logical deduction, but as a reasonable guess, a fertile guess from which observable consequences may be derived so that this particular hypothesis can be subjected to experimental testing. Thus, the principle of relativity is seen, not as a revolutionary new step, but as a natural, indeed an almost obvious, completion of Newton's work."

QUESTIONS

1. Distinguish between sound and light as to their value as synchronizing signals. Is there a lack of analogy?

2. If the limiting speed of signals in classical physics were c rather than infinity, would simultaneity be an absolute concept or a relative concept in classical physics?

3. Give an example from classical physics in which the motion of a clock affects its rate, that is, the way it runs. (The magnitude of the effect may depend on the detailed nature of the clock.)

4. Explain how the result of the Michelson-Morley experiment was put into our definition (procedure) of simultaneity (for synchronizing clocks).

5. According to Eqs. 2-4 and 2-5, each inertial observer finds the center of the spherical electromagnetic wave to be at his own origin at all times, even when the origins do not coincide. How is this result related to our procedure for synchronizing clocks?

6. How can we justify excluding the negative roots in solving for the coefficients a_{11} and a_{44} in Section 2-2?

7. What assumptions, other than the relativity principle and the constancy of c, were made in deducing the Lorentz transformation equations?

8. In our deduction of the length contraction, we arrive at the same result that was proposed by Lorentz. Why then did we reject the Lorentz length contraction hypothesis; that is, in what way do our assumptions differ from those of Lorentz?

9. Two observers, one at rest in S and one at rest in S', each carry a meter stick oriented parallel to their relative motion. *Each* observer finds on measurement that the *other* observer's meter stick is shorter than his meter stick. Explain this apparent paradox. (*Hint.* Compare the following situation. Harry waves good-bye to Walter, in the rear of a station wagon driving away from Harry. Harry says that Walter gets smaller. Walter says that Harry gets smaller. Are they measuring the same thing?)

10. Although in relativity (where motion is relative and not absolute) we find that "moving clocks run slow," this effect has nothing to do with motion altering the way a clock works. What does it have to do with?

11. In time dilation, what is dilated? Would "time retardation" be a better term?

12. Is it true that two events which occur at the same place and at the same time for one observer will be simultaneous for all observers? Explain.

13. If an event A *precedes* an event B at the *same* point in one frame of reference, will A precede B in all other inertial reference frames? Will they occur at the same point in any other inertial frame? Will the time interval between the events be the same in any other inertial frame? Explain.

14. If two events are simultaneous but separated in space in frame S, will they be simultaneous in any other frame S'? Will their space separation be the same in any other frame? Explain.

15. Explain, using the velocity addition theorem of relativity, how we can account for the result of the Michelson-Morley experiment and the double-star observations.

16. Equation 2-17 for the relativistic addition of parallel velocities holds whether u' and v are positive or negative, although our examples considered only positive quantities. Modify an example to include a negative value for u' or v and show that the physical conclusions are unchanged.

17. Compare the results obtained for length- and time-interval measurements by observers in frames whose relative velocity is c. In what sense, from this point of view, does c become a limiting velocity?

18. In Example 5, what would happen if $v_w = -c/n$?

19. Consider a spherical light wavefront spreading out from a source. As seen by the source, what is the *difference in velocity* of portions of the wavefront traveling in opposite directions? What is the *relative velocity* of one portion of the wavefront with respect to the other portion?

20. The sweep rate of the tail of a comet can exceed the speed of light. Explain this phenomenon and show that there is no contradiction with relativity.

21. Imagine a source of light emitting radiation (photons) uniformly in all directions in S'. In S, the radiation will be concentrated in the forward direction for high values of v. Explain, qualitatively (see Problem 37).

22. List several experimental results not predicted or explained by classical physics which are predicted or explained by special relativity theory.

23. Is everything relative according to relativity theory or are there any invariant things permitted by the theory? That is, are there any things which appear to be the same for all observers? If so, name some of them.

24. Why *is* Einstein's theory called the theory of relativity? Would some other name characterize it better?

25. Is the classical concept of an incompressible fluid valid in relativity? Explain.

26. We have stressed the utility of relativity at high speeds. Relativity is also useful in cosmology, where great distances and long time intervals are involved. Show, from the form of the Lorentz transformation equations, why this is so.

PROBLEMS

1. Show that Eqs. 2-6 for a_{44}, a_{11}, and a_{41} are solutions to the equations preceding them.

★**2.** Use the principle of relativity to prove that the two middle equations of Eqs. 2-1 reduce to $y' = y$ and $z' = z$. (See Ref. 3.)

3. Derive Eqs. 2-8 directly from Eqs. 2-7.

4. Suppose that an event occurs in S at $x = 100$ km, $y = 10$ km, $z = 1.0$ km at $t = 2.0 \times 10^{-4}$ seconds. Let S' move relative to S at $0.95c$ along the common x-x' axis, the origins coinciding at $t' = t = 0$. What are the coordinates x', y', z', and t' of this event in S'? Check the answer by using the inverse transformation to obtain the original data.

5. At what speed v will the Galilean and Lorentz expressions for x differ by 0.10 percent? By 1 percent? By 10 percent?

6. Two events, one at position x_1, y_1, z_1 and another at a different position $x_2, y_2,$ z_2 occur at the *same time t* according to observer S. (*a*) Do these events appear to be simultaneous to an observer in S' who moves relative to S at speed v? (*b*) If not, what is the time interval he measures between occurrences of these events? (*c*) How is this time interval affected as $v \to 0$? As the separation between events goes to zero?

7. A cart moves on a track with a constant velocity v (See Fig. 2-10). A and B are on the ends of the cart and observers C and D are stationed along the track. We define event AC as the occurrence of A passing C, and the others similarly. (*a*) Of the four events BD, BC, AD, AC, which are useful for measuring the rate of a clock carried by A for observers along the track? (*b*) Let Δt be the time interval between these two events for observers along the track. What time interval does the moving clock show? (*c*) Suppose that the events BC and AD are simultaneous in the track reference frame. Are they simultaneous in the cart's reference frame? If not, which is earlier?

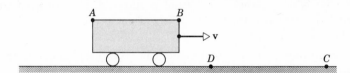

Figure 2-10. Problems 7 and 8.

8. A cart moves on a track with constant velocity, as in Problem 7. Event AD is simultaneous with BC in the track frame. (*a*) The track observers set out to measure the length of the cart AB. They can do so either by using the events BD and AD and working through time measurements or by using the events BC and AC. In either case, the observers in the cart are not apt to regard these results as valid. Explain why for each case. (*b*) Suppose that the observers in the cart seek to measure the distance DC by making simultaneous marks on a long meter stick. Where (relative to A and B) would the observer, E, be situated such that AD is simultaneous with EC in the cart frame? Explain why in terms of synchronization. Can you see why there is a length contraction?

9. As seen from inertial system S an event occurs at point A on the x-axis and then 10^{-6} sec later an event occurs at point B further out on the x-axis. A and B are 600 m apart as seen from S. (*a*) Does there exist another inertial system S', moving with speed less than c parallel to the x-axis, such that the two events appear simultaneous as seen from S'? If so, what is the magnitude and direction of the velocity of S' with respect to S? What is the separation of events A and B according to S'? (*b*) Repeat part a for the case where A and B are only 100 m apart as seen from S.

10. Inertial frame S' moves with respect to inertial frame S in the x-direction at a speed $v = 0.6c$ and the origins coincide when $t = t' = 0$ there. Two events are recorded. In frame S event 1 occurs at the origin at $t = 0$ and event 2 occurs on the x-axis at $x = 3000$ m and $t = 4 \times 10^{-6}$ sec. (*a*) Find the times of the events as registered on S' clocks. (*b*) Explain the different observed time order of the events.

11. What is the proper time interval between the occurrence of two events: (*a*) if in some inertial frame the events are separated by 10^9 m and occur 5 sec apart? (*b*) If $\ldots 7.5 \times 10^8$ m and occur 2.5 sec apart? (*c*) If $\ldots 5 \times 10^8$ m and occurs 1.5 sec apart?

12. In our physical derivation of the length contraction [Section 2-4(*c*)] we assumed that the time dilation was given. In a similar manner derive the time dilation for longitudinal light paths, assuming instead that the length contraction is given.

13. We could define the length of a moving rod as the product of its velocity by the time interval between the instant that one end point of the rod passes a fixed marker and the instant that the other end point passes the same marker. Show that this definition also leads to the length contraction result of Eq. 2-10. (*Hint.* Let the rod be at rest in the primed frame and note that the marker is fixed at one position in the unprimed frame.)

14. An airplane 40.0 m in length in its rest system is moving at a uniform velocity with respect to earth at a speed of 630 m/sec. (*a*) By what fraction of its rest length will it appear to be shortened to an observer on earth? (*b*) How long would it take by earth clocks for the airplane's clock to fall behind by one microsecond? (Assume that special relativity only applies).

15. The rest radius of the earth may be taken as 6400 km and its orbital speed about the sun as 30 km/sec. By how much would the earth's diameter appear to be shortened to an observer on the sun, due to the earth's orbital motion?

16. A 100-Mev electron, for which $\beta = 0.999975 = 1 - 0.000025$, moves along the axis of an evacuated tube which has a length l' of 3.00 m, as measured by a laboratory observer S' with respect to whom the tube is at rest. An observer S moving with the electron would see the tube moving past at a speed v. What length would observer S measure for this tube? (*Hint.* Use binomial expansion.)

17. The length of a spaceship is measured to be exactly half its proper length. (*a*) What is the speed of the spaceship relative to the observer's frame? (*b*) What is the dilation of the spaceship's unit time?

18. The radius of our galaxy is 3×10^{20} m, or about 3×10^4 light-years. (*a*) Can a person, in principle, travel from the center to the edge of our galaxy in a normal lifetime? Explain, using either time-dilation or length-contraction arguments. (*b*) What constant velocity would he need to make the trip in 30 years (proper time)?

19. Two spaceships, each of proper length 100 m, pass near one another heading in opposite directions. If an astronaut at the front of one ship measures a time interval of 2.50×10^{-6} sec for the second ship to pass him, then (*a*) what is the relative velocity of the spaceships? (*b*) What time interval is measured on the first ship for the front of the second ship to pass from the front to the back of the first ship?

20. Consider a universe in which the speed of light $c = 100$ mi/hr. A Lincoln Continental traveling at a speed v relative to a fixed radar speed trap overtakes a Volkswagen traveling at the speed limit of 50 mi/hr $= c/2$. The Lincoln's speed is such that its length is measured by the fixed observer to be the same as that of the Volkswagen. By how much is the Lincoln exceeding the speed limit? The proper length of the Lincoln is twice that of the Volkswagen.

21. (*a*) If the average (proper) lifetime of a μ-meson is 2.3×10^{-6} sec, what average distance would it travel in vacuum before dying as measured in reference frames in which its velocity is $0.00c$, $0.60c$, $0.90c$, and $0.99c$? (*b*) Compare each of these distances with the distance the meson sees itself traveling through.

22. In the target area of an accelerator laboratory there is a straight evacuated tube 300 m long. A momentary burst of one million radioactive particles enters at one end of the tube moving at a speed of $0.8c$. Half of them arrive at the other end without having decayed. (*a*) How long is the tube as measured in the rest frame of the particles? (*b*) What is the half-life of the particles (the time during which half the particles initially present have decayed) measured in this same frame? (*c*) With what speed is the tube measured to move in the rest frame of the particles?

23. A π^+ meson is created in a high-energy collision of a primary cosmic-ray particle in the earth's atmosphere 200 km above sea level. It descends vertically at a speed of 0.99c and disintegrates, in its proper frame, 2.5 $\times$ 10^{-8} sec after its creation. At what altitude above sea level is it observed from earth to disintegrate?

24. The mean lifetime of μ-mesons stopped in a lead block in the laboratory is measured to be 2.3 $\times$ 10^{-6} sec. The mean lifetime of high-speed μ-mesons in a burst of cosmic rays observed from the earth is measured to be 1.6 $\times$ 10^{-5} sec. Find the speed of these cosmic-ray μ-mesons.

25. Laboratory experiments on μ-mesons at rest show that they have a (proper) average lifetime of about 2.3 $\times$ 10^{-6} sec. Such μ-mesons are produced high in the earth's atmosphere by cosmic-ray reactions and travel at a speed 0.99c relative to the earth a distance of from 4000 to 13000 m after formation before decaying. (*a*) Show that the average distance a μ-meson can travel before decaying is much less than even the shorter distance of 4000 m, if its lifetime in flight is only 2.3 $\times$ 10^{-6} sec. (*b*) Explain the consistency of the observations on length traveled and lifetime by computing the lifetime of a μ-meson in flight as measured by a ground observer. (*c*) Explain the consistency by computing the length traveled as seen by an observer at rest on the meson in its flight through the atmosphere.

26. (*a*) Derive Eq. 2-18 in the same way in which Eq. 2-19 was derived. (*b*) Derive Eq. 2-19 directly, rather than by taking the inverse of u'_y.

★**27.** In Fig. 2-11, A and B are the points of intersection of the x-axes (stationary rod) and an inclined rod (moving rod) at two different times. The inclined rod is moving in the +y-direction (without turning) with a speed v. (*a*) Show that the point of intersection of the rods has a speed $u = v \cot \theta$ to the left. (*b*) Let $\theta = 30°$ and $v = \frac{2}{3}c$. Show that u then exceeds c and explain why no contradiction with relativity exists.

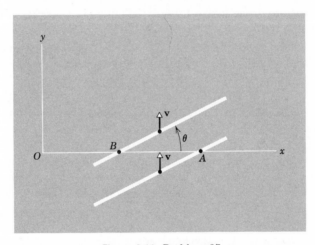

Figure 2-11. Problem 27.

28. An observer in an inertial system S reports that two missiles are moving parallel to one another on a straight line path, one with a speed 0.9c and the other with a speed 0.7c. Find the speed of one missile with respect to the other.

29. One cosmic-ray particle approaches the earth along its axis with a velocity 0.8c toward the North Pole and another with a velocity 0.6c toward the South Pole. What is the relative speed of approach of one particle with respect to the other? (*Hint.* It is useful to consider the earth and one of the particles as the two inertial systems.)

30. Suppose that a particle moves parallel to the x-x' axis, that $v = $ 25,000 mi/hr and $u'_x = $ 25,000 mi/hr. What percent error is made in using the Galilean rather than the Lorentz equation to calculate u_x? The speed of light is 6.7 $\times$ 10^8 mi/hr.

★31. Although in special relativity the frames are inertial (unaccelerated), the objects whose motions we study may be accelerating with respect to such frames. We can obtain the relativistic *acceleration* transformation equations by time differentiation of the velocity transformation equations. (*a*) Show that, with $a_x = du_x/dt$ and $a'_x = du'_x/dt'$ as the x and x' components of the acceleration,

$$a'_x = a_x \frac{(1 - v^2/c^2)^{3/2}}{(1 - u_x v/c^2)^3}.$$

[*Hint.* $du_x/dt' = (du_x/dt)(dt/dt').$]

(*b*) Show that this relativistic result reduces to the classical result ($a'_x = a_x$) when u and v are small compared to c.

32. Consider a radioactive nucleus moving with uniform velocity $0.05c$ relative to the laboratory. (*a*) The nucleus decays by emitting an electron with a speed $0.8c$ along the direction of motion (the common x-x' axis). Find the velocity (magnitude and direction) of the electron in the lab frame, S. (*b*) The nucleus decays by emitting an electron with speed $0.8c$ along the positive y'-axis. Find the velocity (magnitude and direction) of the electron in the lab frame. (*c*) The nucleus decays by emitting an electron with a speed $0.8c$ along the positive y-axis (i.e., perpendicular to the original motion of the nucleus in the lab frame). Find the speed of the electron and the direction of emission in the original rest frame of the nucleus, S'.

★33. Suppose that event A causes event B in frame S, the effect being propagated with a speed *greater than c*. Show, using the velocity addition theorem, that there exists an inertial frame S', which moves relative to S with a velocity less than c, in which the order of these events would be reversed. Hence, if concepts of cause and effect are to be preserved, it is impossible to send signals with a speed greater than that of light.

34. A stick at rest in S has a length L and is inclined at an angle θ to the x-axis (see Fig. 2-12). Find its length L' and angle of inclination θ' to the x'-axis as measured by an observer in S' moving at a speed v relative to S along the x-x' axes.

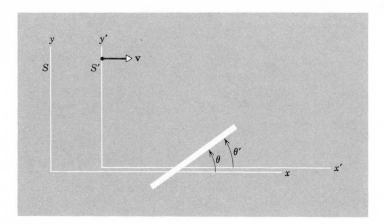

Figure 2-12. Problem 34.

★35. Show that the exact relativistic aberration formula, Eq. 2-27*a*, can be derived from the velocity transformation equations, Eqs. 2-18 and 2-19. (*Hint.* Let a source S', such as an atom, emit light at an angle θ' to the x'-axis of its own rest frame as in Fig. 2-8, and find the emitting angle θ in the S-frame.)

★**36.** An object moves with speed u at an angle θ to the x-axis in system S. A second system S' moves with speed v relative to S along x. What speed u' and angle θ' will the object appear to have to an observer in S'?

37. Imagine a source of light emitting radiation uniformly in all directions in rest-frame S'. Find the distribution of radiation in the laboratory frame S in which the source moves at a speed $\frac{4}{5}c$. (*Hint.* Find the corresponding angle θ for $\theta' = 0, 30,$ 60, 90, 120, 150, and 180°. A polar graph plot of the data would be helpful.) Can you guess why this phenomena is often referred to as the "headlight effect"?

38. A, on earth, signals with a flashlight every six minutes. B is on a space station that is stationary with respect to the earth. C is on a rocket traveling from A to B with a constant velocity of $0.6c$ relative to A (see Fig. 2-13). (a) At what intervals does B receive the signals from A? (b) At what intervals does C receive signals from A? (c) If C flashes a light using intervals equal to those he received from A, at what intervals does B receive C's flashes?

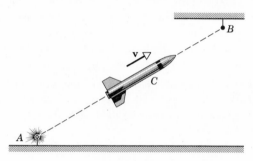

Figure 2-13. Problem 38.

39. A radar transmitter (T) is fixed to a system S_2 which is moving to the right with speed v relative to system S_1 (see Fig. 2-14). A timer in S_2, having a period τ_0 (measured in S_2) causes transmitter T to emit radar pulses, which travel at the speed of light, and are received by R, a receiver fixed to S_1. (a) What would be the period (τ) of the timer relative to observers A and B, spaced a distance $v\tau$ apart? (b) Show that the receiver R would observe the time interval between pulses arriving from S_2 not as τ or as τ_0 but as $\tau_R = \tau_0 \sqrt{(c + v)/(c - v)}$. ($c$) Explain why the observer at R measures a different period for the transmitter than do observers A and B who are in the same reference frame. (*Hint.* Compare the events measured by R to the events measured by A and B. What is meant by the proper time in each case?)

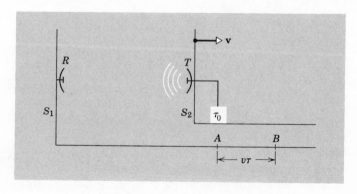

Figure 2-14. Problem 39.

40. In the spectrum of Quasar 3C9, some of the familiar hydrogen lines appear, but they are shifted so far toward the red that their frequencies are observed to be only one-third as large as that observed in a hydrogen rest frame. (*a*) Show that the classical Doppler equation gives a velocity of recession greater than *c*. (*b*) Assuming that the relative motion of 3C9 and the earth is entirely one of recession, find the recession speed predicted by the relativistic Doppler equation.

★41. In the case of wave propagation in a medium, the Doppler shifts for the case of source moving through medium and observer moving through medium directly at one another are different, whereas for light in vacuo the two situations are equivalent. Show that if we take the geometric mean of the two former results, we get exactly the relativistic Doppler shift (see Secs. 20-7 and 40-5 of Ref. 10).

42. A rocketship is receding from the earth at a speed of 0.2*c*. A light in the rocketship ($\lambda \cong 4500$ Å) appears blue to passengers on the ship. What color would it appear to be to an observer on the earth?

43. A rocketship moving away from the earth at a speed $v = (12/13)c$ reports back by sending waves of frequency 100 megacycles/sec as measured in the frame of the rocket ship. At what frequency are the signals received on earth?

44. Give the wavelength shifts in the relativistic longitudinal Doppler effect for the sodium D_1 line (5896 Å) for source and observer approaching at relative velocities of 0.1*c*, 0.4*c*, 0.8*c*. Is the classical (first-order) result a good approximation?

45. Give the wavelength shift in the relativistic Doppler effect for the 6563 Å H_α line emitted by a star receding from the earth with a relative velocity $10^{-3}c$, $10^{-2}c$, and $10^{-1}c$. Is the classical (first-order) result a good approximation?

46. Give the wavelength shift, if any, in the Doppler effect for the sodium D_2 line (5890 Å) emitted from a source moving in a circle with constant speed 0.1*c* measured by an observer fixed at the center of the circle.

REFERENCES

1. *Am. J. Phys.*, January 1963, p. 47.

2. Robert Resnick, *Introduction to Special Relativity,* John Wiley, New York, 1968. See Suppl. Topic A.

3. Ibid. See pp. 58–59.

4. David H. Frisch and James H. Smith, "Measurement of Relativistic Time Dilation Using μ-Mesons," *Am. J. Phys.*, **31**, 342 (1963); and the related film "Time Dilation—An Experiment with μ-Mesons," Educational Services, Inc., Watertown, Mass.

5. V. T. Weisskopf, "The Visual Appearance of Rapidly Moving Objects," *Physics Today* 13(9) (September 1960).

6. N. C. McGill, "The Apparent Shape of Rapidly Moving Objects in Special Relativity," *Contemporary Physics,* **9**, No. 1 33–48, January, 1968.

7. G. D. Scott and H. J. van Driel, "Geometric Appearances at Relativistic Speeds," *Am. J. Phys.*, **38**, 971 (1970).

8. Milton A. Rothman, "Things that go Faster than Light," *Scientific American,* **203**, 142 (July 1960).

9. H. E. Ives and G. R. Stilwell, *J. Opt. Soc. Am.*, **28**, 215 (1938); and **31**, 369 (1941).

10. D. Halliday and R. Resnick, *Physics,* John Wiley and Sons, 1966, p. 1008.

11. Mandelberg and Witten, *J. Opt. Soc. Am.*, **52**, 529 (1962).

12. G. Otting, *Phys. Z.*, **40**, 681 (1939).

13. Walter Kündig, *Phys. Rev.*, **129**, 2371 (1963).

14. H. Bondi, "The Teaching of Special Relativity," *Physics Education,* **1**, (4), 223 (1966).

CHAPTER 3

Relativistic Dynamics

3.1 Mechanics and Relativity

In Chapter One we saw that experiment forced us to the conclusion that the Galilean transformations had to be replaced and the basic laws of mechanics, which were consistent with those transformations, needed to be modified. In Chapter Two we obtained the new transformation equations, the Lorentz transformations, and examined their implications for kinematical phenomena. Now we must consider dynamic phenomena and find how to modify the laws of classical mechanics so that the new mechanics is consistent with relativity.

Basically, classical Newtonian mechanics is inconsistent with relativity because its laws are invariant under a Galilean transformation and *not* under a Lorentz transformation. This formal result is plausible, as well, for other considerations. For example, in Newtonian mechanics a force can accelerate a particle to indefinite speeds, whereas in relativity the limiting speed is c. We seek a new law of motion that is consistent with relativity. When we obtain such a law of motion, we must also insure that it reduces to the Newtonian form as $v/c \rightarrow 0$ since, in the domain where $v/c \ll 1$, Newton's laws are consistent with experiment. Thus, the relativistic law of motion will be a generalization of the classical one.

We shall proceed by studying collisions. Here we assume that the interaction between particles takes place only during an infinitesimally short time interval in which the particles have negligible separation (i.e., the range of forces is short compared to the dimensions of the system). During the collision the particles are accelerated, but before and after the interaction there is no acceleration. The laws of conservation of momentum and energy are valid classically during this interaction. If we require that these conservation laws also be valid relativistically (i.e., invariant under a Lorentz transformation) and hence that they be general laws of physics, we must modify them from the classical form in such a way that they also reduce to the classical form as $v/c \rightarrow 0$. In this way, we shall obtain the relativistic law of motion.

3.2 The Need to Redefine Momentum

The first thing we wish to show is that if we want to find a quantity like momentum (for which there is a conservation law in classical physics) that is also subject to a conservation law in relativity, then we cannot use the same expression for momentum as the classical one. We must, instead, redefine momentum in order that a law of conservation of momentum in collisions be invariant under a Lorentz transformation.

Let us first analyze a special elastic collision between two identical bodies as seen by different inertial observers, S and S', according to Newtonian mechanics. We choose the collision (Fig. 3-1) to be highly symmetrical in S': the bodies, say A and B, have initial velocities that are equal in magnitude but opposite in direction, the total momentum being zero. That is, $\mathbf{u}'_{yA} = -\mathbf{u}'_{yB}$ and $\mathbf{u}'_{xA} = -\mathbf{u}'_{xB}$. Since the collision is elastic, the final velocities have the same magnitude as the initial velocities, the total momentum after colli- sion remaining zero. We have $\mathbf{u}'_{yA} = -\mathbf{U}'_{yA} = \mathbf{U}'_{yB} = -\mathbf{u}'_{yB}$ and $\mathbf{u}'_{xA} = \mathbf{U}'_{xA} = -\mathbf{U}'_{xB} = -\mathbf{u}'_{xB}$. That is, observer S' notes that the y'-components of velocity for the bodies simply reverse their signs during the collision, the x'-components remaining unchanged.

As seen by observer S, the reference frame S' is moving to the right with a speed v. We deliberately choose

$$\mathbf{v} = \mathbf{u}'_{xB} = -\mathbf{u}'_{xA} \tag{3-1}$$

so that the body A has no x-component of motion in frame S (see Fig. 3-1b). The y-components of velocity should be unaffected by the transformation, according to Newtonian mechanics, and momentum should still be conserved in the collision as viewed by S. That is, $\mathbf{u}_{yA} = \mathbf{u}'_{yA}$, $\mathbf{u}_{yB} = \mathbf{u}'_{yB}$, $\mathbf{u}_{yA} = -\mathbf{U}_{yA}$ and $\mathbf{u}_{yB} = -\mathbf{U}_{yB}$. The momentum lost by body A, $2mu_{yA}$, equals that gained by body B, $2mu_{yB}$, so that in magnitude

$$2mu_{yA} = 2mu_{yB} \tag{3-2}$$

and, because the bodies have identical mass m, we conclude that

$$u_{yA} = u_{yB}. \tag{3-3}$$

These are the Newtonian results.

Now, let us see whether these results are consistent with the Lorentz transformations. They are not, for they contradict the relativistic velocity transformations. If we use the equations in Table 2-2 we find that relativity requires, for body B,

$$u'_{yB} = \frac{u_{yB}\sqrt{1 - \beta^2}}{1 - u_{xB}v/c^2} \tag{3-4}$$

whereas for body A, for which $u_{xA} = 0$,

$$u'_{yA} = u_{yA}\sqrt{1 - \beta^2}. \tag{3-5}$$

Hence, the y-components of velocity *are* affected by the relativistic trans- formations. For one thing, they do not have the same values in one frame as in the other, but, more important, if they are equal to one another in

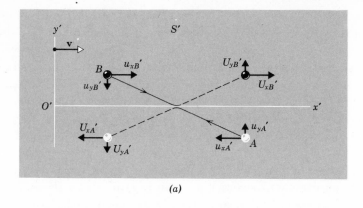

(a)

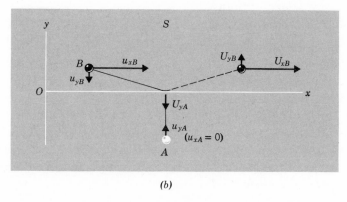

(b)

Figure 3-1. A particular elastic collision as viewed by (a) observer S' and (b) observer S. Here, small letters (u) refer to before the collision, capital letters (U) refer to after the collision. The subscripts (A and B) denote the particle and (x and y) the component. The values in S' are primed, those in S are not.

magnitude in one frame, they are not necessarily equal to one another in the other frame. In fact, assuming as before that $u_{yB}' = u_{yA}'$, we find by combining and rearranging Eqs. 3-4 and 3-5, that

$$u_{yA} = u_{yB}\frac{1}{1 - u_{xB}v/c^2}, \tag{3-6}$$

in contradiction to the Newtonian result, Eq. 3-3. Hence, the *changes* in the y-component velocities have different magnitudes in one frame than in the other during the collision. The result is that, if we compute momentum according to the classical formulas $\mathbf{p} = m\mathbf{u}$ and $\mathbf{p}' = m\mathbf{u}'$, then when momentum is conserved in a collision in one frame it is not conserved in the other frame.

This result contradicts the basic postulate of special relativity that the laws of physics are the same in all inertial systems. If the conservation of momentum in collisions is to be a law of physics, then the classical definition of momentum cannot be correct in general. We notice that the disagreement between Eqs. 3-3 and 3-6 becomes trivial when $u_{xB} \ll c$ and $v \ll c$, so that it is at high speeds that the Newtonian formulation of the momentum conservation law breaks down. We need a generalization of the definition of momentum, therefore, that reduces to the classical result at low speeds.

In the next section, we shall show that it is possible to preserve the *form* of the classical definition of the momentum of a particle, $\mathbf{p} = m\mathbf{u}$, where $\mathbf{p}$ is the momentum, m the mass, and $\mathbf{u}$ the velocity of a particle, and also to preserve the classical law of the conservation of momentum of a system of interacting particles, providing that we modify the classical concept of mass. We need to let the mass of a particle be a function of its speed u, that is, $m = m_0/\sqrt{1 - u^2/c^2}$, where m_0 is the classical mass and m is the relativistic mass of the particle. Clearly, as u/c tends to zero, m tends to m_0. The relativistic momentum then becomes $\mathbf{p} = m\mathbf{u} = m_0\mathbf{u}/\sqrt{1 - \beta^2}$ and reduces to the classical expression $\mathbf{p} = m_0\mathbf{u}$ as $\beta \to 0$. Let us now deduce these results.

3.3 Relativistic Momentum

In Eq. 3-2, based on momentum conservation, we assumed that the mass m was the same for each body, and, in this way, we were led to the (incorrect) result that the y-component velocities had equal magnitude. True, the bodies were identical when placed side by side at rest. However, since the measured length of a rod and the measured rate of a clock are affected by the motion of the rod or the clock relative to the observer, it may be that the measured mass of a body also depends on its motion with respect to the observer. In that case the *form* of the Newtonian momentum still could be correct so that, for example, we could rewrite Eq. 3-2 as

$$2m_A u_{yA} = 2m_B u_{yB}. \tag{3-7}$$

The masses are now labelled as m_A and m_B, however, to suggest that they may have different values.

Bodies A and B, in Fig. 3-1*b*, do travel at different speeds in the S-frame and, if we accept the relativistic result (Eq. 3-6) for the speeds, we obtain

$$m_B = m_A \frac{u_{yA}}{u_{yB}} = \frac{m_A}{1 - u_{xB}v/c^2} \tag{3-8}$$

by combining Eqs. 3-6 and 3-7. Hence, the *relativistic masses,* m_A and m_B, are *not* equal if the relativistic conservation of momentum law is to have the same form as the Newtonian law. It remains to find how the relativistic mass must vary with the speed.

We can simplify Eq. 3-8 by eliminating v. Recall that $v = u'_{xB}$ (Eq. 3-1) and that u'_{xB} is related to u_{xB} by the Lorentz velocity transformation (Table 2-2)

$$u'_{xB}(= v) = \frac{u_{xB} - v}{1 - u_{xB}v/c^2}.$$

Solving for v, we get

$$v = \frac{c^2}{u_{xB}}(1 - \sqrt{1 - (u_{xB}/c)^2}).$$

If we substitute this expression for v into Eq. 3-8 we obtain

$$m_B = \frac{m_A}{\sqrt{1 - (u_{xB}/c)^2}}.$$ (3-9)

We can find how the relativistic mass of either particle varies with the speed in a simple manner by considering a special case of the collision in which the y-y' velocity components are made to approach zero. Then, the particles' speeds will be identical to the magnitude of their respective x-component velocities. This is illustrated in Fig. 3-2a and 3-2b. Observer S' simply sees two bodies moving past each other making a grazing collision; observer S sees body A at rest and body B moving past it, at a speed u_{xB}, again making a grazing collision. Equation 3-9 must apply to this grazing collision as well as to others because we put no restriction on the value of u'_y in deriving it.

Since body A is at rest in S its mass m_A must be the ordinary Newtonian mass which we now call the *rest mass* and denote by m_0. This is the same as the mass of body B when body B is at rest, the two bodies being identical. However, in S, body B is moving with a speed u_{xB}, which we can simply call u; its mass m_B, which we can call the *relativistic mass* and denote by m, will not be m_0. From Eq. 3-9 we obtain

$$m = \frac{m_0}{\sqrt{1 - u^2/c^2}}$$ (3-10)

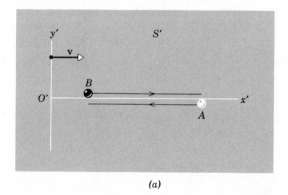

(a)

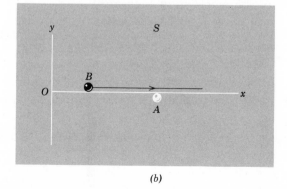

(b)

Figure 3-2. The same collision as in Fig. 3-1 for the limiting case in which $u_{yA'} = u_{yB'} = 0$.

which tells us how the relativistic mass m of a body moving at a speed u varies with u. We see at once that when $u = 0$, the body then being at rest, we obtain $m = m_0$, the rest mass. More generally, as $u/c \to 0$, we find $m \to m_0$, which is the Newtonian limit of the more general expression for the relativistic mass m.

Hence, if we want to preserve the *form* of the classical momentum conservation law while requiring that the law be relativistically invariant, we must define the mass of a moving body by Eq. 3-10. That is, momentum still has the form $m\mathbf{u}$, but mass is defined as $m = m_0/\sqrt{1 - u^2/c^2}$. Note that u is the speed of the body relative to S, which we can regard as the laboratory frame, and that u has no connection necessarily with changing reference frames. By accepting Eq. 3-10 as our definition of the mass of a moving body, we implicitly assume that the mass of a body does not depend on its acceleration relative to the reference frame, although it does depend on its speed. Mass remains a scalar quantity in the sense that its value is independent of the *direction* of the velocity of the body. The rest mass m_0 is often called the *proper mass,* for it is the mass of the body measured, like proper length and proper time, in the inertial frame in which the body is at rest.

We have presented above a derivation of an expression for relativistic momentum which obviously centers around a very special case. For example, the velocity of the particle (B) is parallel to the relative S-S' velocity and the derivation depended only upon invoking conservation of momentum in the y-direction. Such a derivation enables us to make an educated guess as to what the general result may be. We have avoided rather involved general derivations which, however, lead to exactly the same results. When the general case is done, u becomes the absolute value of the velocity of the particle; that is, $u^2 = u_x^2 + u_y^2 + u_z^2$.

Hence, to conclude, in order to make the conservation of momentum in collisions a law that is experimentally valid in all reference frames, we must define momentum, not as $m_0\mathbf{u}$, but as

$$\mathbf{p} = \frac{m_0\mathbf{u}}{\sqrt{1 - u^2/c^2}}. \tag{3-11}$$

The components of the momentum then are

$$p_x = \frac{m_0 u_x}{\sqrt{1 - u^2/c^2}} \qquad p_y = \frac{m_0 u_y}{\sqrt{1 - u^2/c^2}} \qquad p_z = \frac{m_0 u_z}{\sqrt{1 - u^2/c^2}} \tag{3-12}$$

which we write out explicitly to emphasize that the magnitude u of the total velocity appears in the denominator of each component equation.

▶ **Example 1.** For what value of $u/c \, (= \beta)$ will the relativistic mass of a particle exceed its rest mass by a given fraction f?

From Eq. 3-10 we have

$$f = \frac{m - m_0}{m_0} = \frac{m}{m_0} - 1 = \frac{1}{\sqrt{1 - \beta^2}} - 1$$

which, solved for β, is

$$\beta = \frac{\sqrt{f(2 + f)}}{1 + f}$$

The table below shows some computed values, which hold for all particles regardless of their rest mass.

f		β
0.001	(0.1 percent)	0.045
0.01		0.14
0.1		0.42
1	(100 percent)	0.87
10		0.994
100		0.999 ◀

3.4   The Relativistic Force Law and the Dynamics of a Single Particle

Newton's second law must now be generalized to

$$\mathbf{F} = \frac{d}{dt}(\mathbf{p}) = \frac{d}{dt}\left(\frac{m_0\mathbf{u}}{\sqrt{1 - u^2/c^2}}\right) \tag{3-13}$$

in relativistic mechanics. When the law is written in this form we can immediately deduce the law of the conservation of relativistic momentum from it; when $\mathbf{F}$ is zero, $\mathbf{p} = m_0\mathbf{u}/\sqrt{1 - u^2/c^2}$ must be a constant. In the absence of external forces, the momentum is conserved. Notice that this new form of the law, Eq. 3-13, is *not* equivalent to writing

$$F = ma = \left(\frac{m_0}{\sqrt{1 - u^2/c^2}}\right)\left(\frac{du}{dt}\right),$$

in which we simply multiply the acceleration by the relativistic mass.

We find that experiment agrees with Eq. 3-13. When, for example, we investigate the motion of high-speed charged particles, it is found that the equation correctly describing the motion is

$$q(\mathbf{E} + \mathbf{u} \times \mathbf{B}) = \frac{d}{dt}\left(\frac{m_0\mathbf{u}}{\sqrt{1 - u^2/c^2}}\right), \tag{3-14}$$

which agrees with Eq. 3-13. Here, $q(\mathbf{E} + \mathbf{u} \times \mathbf{B})$ is the Lorentz electromagnetic force, in which $\mathbf{E}$ is the electric field, $\mathbf{B}$ is the magnetic field, and $\mathbf{u}$ is the particle velocity, all measured in the same reference frame, and q and m_0 are constants that describe the electrical (charge) and inertial (rest mass) properties of the particle, respectively (see *Physics*, Part II, Sec. 33-2). Notice that the form of the Lorentz force law of classical electromagnetism remains valid relativistically, as we should expect from the discussion of Chapter One.

Later we shall turn to the question of how forces transform from one Lorentz frame to another. For the moment, however, we confine ourselves to one reference frame (the laboratory frame) and develop other concepts in mechanics, such as work and energy, which follow from the relativistic expression for force (Eq. 3-13). We shall confine ourselves to the motion of a single particle. In succeeding sections we shall consider many-particle systems.

In Newtonian mechanics we define the kinetic energy, K, of a particle to be equal to the work done by an external force in increasing the speed of the particle from zero to some value u (see *Physics,* Part I, Sec. 7-5). That is,

$$K = \int_{u=0}^{u=u} \mathbf{F} \cdot d\mathbf{l}$$

where $\mathbf{F} \cdot d\mathbf{l}$ is the work done by the force $\mathbf{F}$ in displacing the particle through $d\mathbf{l}$. For simplicity, we can limit the motion to one dimension, say x, the three-dimensional case being an easy extension (see Problem 8). Then, classically,

$$K = \int_{u=0}^{u=u} F\, dx = \int m_0 \left(\frac{du}{dt}\right) dx = \int m_0\, du \frac{dx}{dt} = m_0 \int_0^u u\, du = \tfrac{1}{2} m_0 u^2.$$

Here we write the particle mass as m_0 to emphasize that, in Newtonian mechanics, we do not regard the mass as varying with the speed, and we take the force to be $m_0 a = m_0(du/dt)$.

In relativistic mechanics, it proves useful to use a corresponding definition for kinetic energy in which, however, we use the relativistic equation of motion, Eq. 3-13, rather than the Newtonian one. Then, relativistically,

$$K = \int_{u=0}^{u=u} F\, dx = \int \frac{d}{dt}(mu)\, dx = \int d(mu)\frac{dx}{dt}$$

$$= \int (m\, du + u\, dm)\, u = \int_{u=0}^{u=u} (mu\, du + u^2\, dm) \qquad (3\text{-}15)$$

in which both m and u are variables. These quantities are related, furthermore, by Eq. 3-10, $m = m_0/\sqrt{1 - u^2/c^2}$, which we can rewrite as

$$m^2 c^2 - m^2 u^2 = m_0^2 c^2.$$

Taking differentials in this equation yields

$$2mc^2\, dm - m^2 2u\, du - u^2 2m\, dm = 0,$$

which, on division by $2m$, can be written also as

$$mu\, du + u^2\, dm = c^2\, dm.$$

The left side of this equation is exactly the integrand of Eq. 3-15. Hence, we can write the relativistic expression for the kinetic energy of a particle as

$$K = \int_{u=0}^{u=u} c^2\, dm = c^2 \int_{m=m_0}^{m=m} dm = mc^2 - m_0 c^2. \qquad (3\text{-}16a)$$

By using Eq. 3-10, we obtain equivalently

$$K = m_0 c^2 \left[\frac{1}{\sqrt{1 - u^2/c^2}} - 1 \right]. \qquad (3\text{-}16b)$$

Also, if we take $mc^2 = E$, where E is called the *total energy* of the particle—a name whose aptness will become clear later—we can express Eqs. 3-16 compactly as

$$E = m_0c^2 + K \qquad (3\text{-}17)$$

in which m_0c^2 is called the *rest energy* of the particle. The rest energy (by definition) is the energy of the particle at rest, when $u = 0$ and $K = 0$. The total energy of the particle (Eq. 3-17) is the sum of its rest energy* and its kinetic energy.

The relativistic expression for K must reduce to the classical result, $\frac{1}{2}m_0u^2$, when $u/c \ll 1$. Let us check this. From

$$K = m_0c^2[(1/\sqrt{1 - u^2/c^2}) - 1]$$
$$= m_0c^2\left[\left(1 - \frac{u^2}{c^2}\right)^{-1/2} - 1\right]$$

the binomial theorem expansion in (u/c) gives

$$K = m_0c^2\left[1 + \frac{1}{2}\left(\frac{u}{c}\right)^2 + \frac{3}{8}\left(\frac{u}{c}\right)^4 + \cdots - 1\right]$$
$$= \tfrac{1}{2}m_0u^2,$$

in which we take only the first two terms in the expansion as significant when $u/c \ll 1$, thereby confirming the Newtonian limit of the relativistic result.

It is interesting to notice also that, as $u \to c$, in Eq. 3-16b, the kinetic energy K tends to infinity. That is, from Eq. 3-15, an infinite amount of work would need to be done on the particle to accelerate it up to the speed of light. Once again we find c playing the role of a limiting velocity. Note also from Eq. 3-16a, which permits us to write $K = (m - m_0)c^2$, that a change in the kinetic energy of a particle is related to a change in its (inertial) mass.

We often seek a connection between the kinetic energy K of a rapidly moving particle and its momentum p. This can be found by eliminating u between Eq. 3-16b and Eq. 3-11. The student can verify (Problem 8) that the result is

$$(K + m_0c^2)^2 = (pc)^2 + (m_0c^2)^2 \qquad (3\text{-}18a)$$

which, with the total energy $E = K + m_0c^2$, can also be written as

$$E^2 = (pc)^2 + (m_0c^2)^2. \qquad (3\text{-}18b)$$

The right triangle of Fig. 3-3 is a useful mnemonic device for remembering Eqs. 3-18.

The relationship between K and p (Eq. 3-18a) should reduce to the Newtonian expression $p = \sqrt{2m_0K}$ for $u/c \ll 1$. To see that it does, let us expand Eq. 3-18a, obtaining

$$K^2 + 2Km_0c^2 = p^2c^2.$$

---

*In classical physics, the energy of a single particle is defined only to within an arbitrary constant. Relativity fixes this arbitrary constant so that the energy of a particle at rest is taken to be $E_0 = m_0c^2$. The physical meaning of this (see Section 3-5) is that even a particle that is not in motion has a *rest* energy, given by m_0c^2.

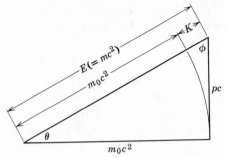

Figure 3-3. A mnemonic device, using a right triangle and the Pythagorean relation, to help in remembering the relations between total energy, E, rest energy m_0c^2 and momentum p. The relation is $E^2 = (pc)^2 + (m_0c^2)^2$. Shown also is the relation, $E = m_0c^2 + K$, between total energy, rest energy, and kinetic energy K. The student can show that $\sin \theta = \beta$ and $\sin \phi = \sqrt{1 - \beta^2}$, where $\beta = u/c$.

When $u/c \ll 1$, the kinetic energy, K, of a moving particle will always be much less than its rest energy, m_0c^2 (see Problem 6). Under these circumstances, the first term on the left above (K^2) can be neglected in comparison with the second term ($2K\,m_0c^2$), and the equation becomes $p = \sqrt{2m_0K}$, as required.

The relativistic expression, Eq. 3-18b, often written as

$$E = c\sqrt{p^2 + m_0^2 c^2},$$ (3-19)

is useful in high-energy physics to calculate the total energy of a particle when its momentum is given, or vice versa. By differentiating Eq. 3-19 with respect to p, we can obtain another useful relation.

$$\frac{dE}{dp} = \frac{pc}{\sqrt{m_0^2 c^2 + p^2}} = \frac{pc^2}{c\sqrt{m_0^2 c^2 + p^2}} = \frac{pc^2}{E}.$$

But with $E = mc^2$ and $\mathbf{p} = m\mathbf{u}$ this reduces to

$$\frac{dE}{dp} = u,$$ (3-20)

a result which, incidentally, is also valid in classical dynamics.

As a final consideration in the relativistic dynamics of a single particle, we look at the acceleration of a particle under the influence of a force. In general, the force is given by

$$\mathbf{F} = d\mathbf{p}/dt = \frac{d}{dt}(m\mathbf{u})$$

or

$$\mathbf{F} = m\frac{d\mathbf{u}}{dt} + \mathbf{u}\frac{dm}{dt}.$$ (3-21)

We know that $m = E/c^2$ so that

$$\frac{dm}{dt} = \frac{1}{c^2}\frac{dE}{dt} = \frac{1}{c^2}\frac{d}{dt}(K + m_0c^2) = \frac{1}{c^2}\frac{dK}{dt}.$$

But
$$\frac{dK}{dt} = \frac{(\mathbf{F} \cdot d\mathbf{l})}{dt} = \mathbf{F} \cdot \frac{d\mathbf{l}}{dt} = \mathbf{F} \cdot \mathbf{u}$$

so that
$$\frac{dm}{dt} = \frac{1}{c^2} \mathbf{F} \cdot \mathbf{u}.$$

We can now substitute this into Eq. 3-21 and obtain

$$\mathbf{F} = m \frac{d\mathbf{u}}{dt} + \frac{\mathbf{u}(\mathbf{F} \cdot \mathbf{u})}{c^2}.$$

The acceleration $\mathbf{a}$ is defined by $\mathbf{a} = d\mathbf{u}/dt$ so that the general expression for acceleration is

$$\mathbf{a} = \frac{d\mathbf{u}}{dt} = \frac{\mathbf{F}}{m} - \frac{\mathbf{u}}{mc^2}(\mathbf{F} \cdot \mathbf{u}). \qquad (3\text{-}22)$$

What this equation tells us at once is that, in general, the acceleration $\mathbf{a}$ is *not* parallel to the force in relativity, since the last term above is in the direction of the velocity $\mathbf{u}$.

▶ **Example 2.** (a) What is the kinetic energy acquired by a particle of charge q starting from rest in a uniform electric field when it falls through an electrostatic potential difference of V_0 volts?

The work done on the charge q by the electric field $\mathbf{E}$ in a displacement $d\mathbf{l}$ is

$$dW = q\mathbf{E} \cdot d\mathbf{l}.$$

Let the uniform field be in the x-direction so that $\mathbf{E} \cdot d\mathbf{l} = E_x\, dx$ and

$$W = \int qE_x\, dx.$$

Now $E_x = -(dV/dx)$, where V is the electrostatic potential, so that

$$W = -\int q \frac{dV}{dx} dx = -q \int dV = -q(V_f - V_i)$$
$$= q(V_i - V_f) = qV_0$$

where V_0 is the difference between the initial potential V_i and the final potential V_f. The kinetic energy acquired by the charge is equal to the work done on it by the field so that

$$K = qV_0. \qquad (3\text{-}23)$$

Notice that we have implicitly assumed that the charge q of the particle is a constant, independent of the particle's motion.

(b) Assume the particle to be an electron and the potential difference to be 10^4 volts. Find the kinetic energy of the electron, its speed, and its mass at the end of the acceleration.

The charge on the electron is $e = -1.602 \times 10^{-19}$ coulomb. The potential difference is now a rise, $V_i - V_f = -10^4$ volts, a negative charge accelerating in a direction opposite to $\mathbf{E}$. Hence, the kinetic energy acquired is

$$K = qV_0 = (-1.602 \times 10^{-19})(-10^4) \text{ joules} = 1.602 \times 10^{-15} \text{ joules}.$$

From Eq. 3-16, $K = mc^2 - m_0c^2$, we obtain

$$\frac{K}{c^2} = (m - m_0)$$

or

$$(1.602 \times 10^{-15} \text{ joules}/8.99 \times 10^{16} \ m^2/\text{sec}^2) = m - m_0 = 1.78 \times 10^{-32} \text{ kg}$$

and, with $m_0 = 9.109 \times 10^{-31}$ kg, we find the mass of the moving electron to be

$$m = (9.109 + 0.178) \times 10^{-31} \text{ kg} = 9.287 \times 10^{-31} \text{ kg}.$$

Notice that $m/m_0 = 1.02$, so that the mass increase due to the motion is about 2 percent of the rest mass.

From Eq. 3-10, $m = m_0/\sqrt{1 - u^2/c^2}$, we have

$$\frac{u^2}{c^2} = \left[1 - \left(\frac{m_0}{m}\right)^2\right] = \left[1 - \left(\frac{9.109}{9.287}\right)^2\right] = 0.038$$

or $u = 0.195c = 5.85 \times 10^7$ m/sec.

The electron acquires a speed of about one-fifth the speed of light.

These are the relativistic predictions. We shall see below that they are confirmed by direct experiment.

Example 3. (a) Show that, in a region in which there is a uniform magnetic field, a charged particle entering at right angles to the field moves in a circle whose radius is proportional to the particle's momentum.

Call the charge of the particle q and its rest mass m_0. Let its velocity be **u.** The force on the particle is then

$$\mathbf{F} = q\mathbf{u} \times \mathbf{B}$$

which is at right angles both to **u** and to **B,** the magnetic field. Hence, from Eq. 3-22, the acceleration,

$$\mathbf{a} = \frac{\mathbf{F}}{m} = \frac{q}{m}\mathbf{u} \times \mathbf{B},$$

is in the same direction as the force. Because the acceleration is always at right angles to the particle's velocity **u,** the speed of the particle is constant and the particle moves in a circle. Let the radius of the circle be r, so that the centripetal acceleration is u^2/r. We equate this to the acceleration obtained from above, $a = quB/m$, and find

$$\frac{quB}{m} = \frac{u^2}{r}$$

or $$r = \frac{mu}{qB} = \frac{p}{qB}. \tag{3-24}$$

Hence, the radius is proportional to the momentum $p(= mu)$.

Notice that both the equation for the acceleration and the equation for the radius (Eq. 3-24) are identical in form to the classical results, but that the rest mass m_0 of the classical formula is replaced by the relativistic mass $m = m_0/\sqrt{1 - u^2/c^2}$.

How would the motion change if the initial velocity of the charged particle had a component parallel to the magnetic field?

(b) Compute the radius, both classically and relativistically, of the path of a 10 Mev electron moving at right angles to a uniform magnetic field of strength 2.0 webers/m^2.

Classically, we have $r = m_0 u/qB$. The classical relation between kinetic energy and momentum is $p = \sqrt{2m_0 K}$ so that

$$p = \sqrt{2m_0 K} = \sqrt{2 \times 9.1 \times 10^{-31} \text{ kg} \times 10 \text{ Mev} \times 1.6 \times 10^{-13} \text{joule/Mev}}$$
$$= 17 \times 10^{-22} \text{ kg-m/sec.}$$

Then

$$r = \frac{m_0 u}{qB} = \frac{p}{qB} = \frac{17 \times 10^{-22}}{1.6 \times 10^{-19} \times 2.0} \text{meter} = 5.3 \times 10^{-3} \text{ meter} = 0.53 \text{ cm.}$$

Relativistically, we have $r = mu/qB$. The relativistic relation between kinetic energy and momentum (Eq. 3-18a) may be written as

$$p = \frac{1}{c} \sqrt{(K + m_0 c^2)^2 - (m_0 c^2)^2}.$$

Here, the rest energy of an electron, $m_0 c^2$, equals 0.51 Mev, so that

$$p = \frac{1}{3 \times 10^8} \sqrt{(10 + 0.51)^2 - (0.51)^2} \frac{\text{Mev-sec}}{\text{meter}} (1.6 \times 10^{-13} \text{ joule/Mev})$$
$$= 5.6 \times 10^{-21} \text{ kg-m/sec.}$$

Then

$$r = \frac{mu}{qB} = \frac{p}{qB} = \frac{5.6 \times 10^{-21}}{1.6 \times 10^{-19} \times 2.0} \text{meter} = 1.8 \times 10^{-2} \text{ meter} = 1.8 \text{ cm.}$$

Experiment bears out the relativistic result (see below). ◀

The first experiments in relativistic dynamics, by Bucherer [1], made use of Eq. 3-24. Electrons (from the β-decay of radioactive particles) enter a velocity selector, which determines the speed of those that emerge, and then enter a uniform magnetic field, where the radius of their circular path can be measured. Bucherer's results are shown in Table 3-1.

The first column gives the measured speeds in terms of the fraction of the speed of light. The second column gives the ratio e/m computed from the measured quantities in Eq. 3-24 as $e/m = u/rB$. It is clear that the value of e/m varies with the speed of the electrons. The third column gives the calculated values of $e/m \sqrt{1 - u^2/c^2} = e/m_0$, which are seen to be constant. The results are consistent with the relativistic relation

$$r = \frac{m_0 u}{qB \sqrt{1 - u^2/c^2}}$$

Table 3-1 BUCHERER'S RESULTS

u/c	$e/m(= u/rB)$ in coul/kg	$\dfrac{e}{m_0}\left(= \dfrac{e}{m\sqrt{1 - u^2/c^2}}\right)$ in coul/kg
(Measured)	(Measured)	(Computed)
0.3173	1.661×10^{11}	1.752×10^{11}
0.3787	1.630×10^{11}	1.761×10^{11}
0.4281	1.590×10^{11}	1.760×10^{11}
0.5154	1.511×10^{11}	1.763×10^{11}
0.6870	1.283×10^{11}	1.767×10^{11}

rather than the classical relation $r = m_0 u / qB$ and can be interpreted as confirming Eq. 3-10, $m = m_0 / \sqrt{1 - u^2/c^2}$, for the variation of mass with speed.* Many similar experiments have since been performed, greatly extending the range of u/c and always resulting in confirmation of the relativistic results (see Fig. 3-4).

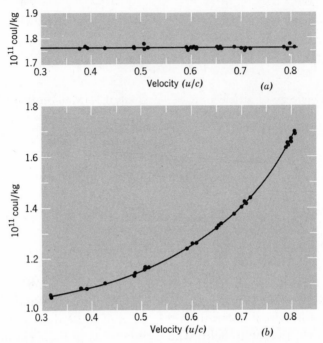

Figure 3-4. Experimental verification of the relativistic mass formula. Experimental points are shown for u/c ranging from 0.32 to 0.82. (*a*) The ratio $e/m_0 = e/m \sqrt{1 - u^2/c^2}$. (*b*) The ratio $m/m_0 = 1/\sqrt{1 - u^2/c^2}$.

The student may properly ask why, in measuring a variation of e/m with speed, we attribute the variation solely to the mass rather than to the charge, for instance, or some other more complicated effect. We might have concluded, for example, that $e = e_0 \sqrt{1 - u^2/c^2}$. Actually, we have implicitly assumed above that the charge on the electron is independent of its speed. This assumption is a direct consequence of relativistic electrodynamics, wherein the charge of a particle is not changed by its motion. That is, charge is an invariant quantity in relativity. This is plausible, as a little thought shows, for otherwise the neutral character of an atom, say, would be upset merely by the motion of the electrons in it. As a clincher, of course, we turn to experiment; we then find that experiment not only verifies relativity theory as a whole but also confirms directly this specific result of the constancy of e (see Refs. 2 and 3 for an analysis of such an experiment).

The relations used in Example 2, above, were tested directly in a recent experiment by Bertozzi [4]. Electrons are accelerated to high speed in the

*These results verify not only that relativity predicts the correct functional form for $m(u)$ but also that the value of the limiting speed (c) is 3×10^{10} cm/sec.

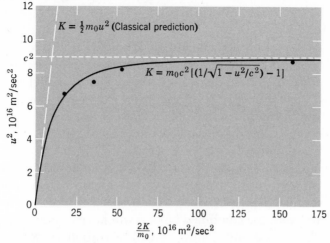

Figure 3-5. Bertozzi's experimental points (dots) are seen to fit the relativistic expression (solid line) rather than the classical expression (dashed line) for kinetic energy K versus u^2.

electric field of a linear accelerator and emerge into a vacuum chamber. Their speed can be measured by determining the time of flight in passing two targets of known separation. As we vary the voltage of the accelerator, we can plot the values of eV, the kinetic energy of the emerging electrons, versus the measured speed u. In the experiment, an independent check was made to confirm the relation $K = eV$. This is accomplished by stopping the electrons in a collector, where the kinetic energy of the absorbed electrons is converted into heat energy which raises the temperature of the collector, and determining the energy released per electron by calorimetry. It is found that the average kinetic energy per electron before impact, measured in this way, agrees with the kinetic energy obtained from eV.

In Fig. 3-5, we show a plot of the results. Here, on the ordinate, is plotted u^2 versus $2K/m_0$, on the absicca. At low energies, the experimental results (solid curve) agree with the classical prediction (dashed curve), $K = \frac{1}{2}m_0 u^2$ (i.e., $2K/m_0 = u^2$). However, as the energy rises, we find that $2K/m_0 > u^2$. In fact, the measured values of u were always less than c, regardless of how high the energy became, so that the experimental curve approaches but never reaches the dotted line corresponding to $u = c$. We see that to attain a given speed we need more kinetic energy than is classically predicted and that, by extrapolation, we would need an infinite energy to accelerate the electron to the speed of light. The experimental curve fits the relativistic prediction of Eq. 3-16b,

$$K = m_0 c^2 \left(\frac{1}{\sqrt{1 - u^2/c^2}} - 1 \right),$$

and can be regarded as another confirmation of the relativistic mass formula of Eq. 3-10, $m = m_0 \sqrt{1 - u^2/c^2}$.

The student should note carefully that the relativistic formula for kinetic energy is *not* $\frac{1}{2}mu^2$; this shows the danger, mentioned earlier, in assuming that we can simply substitute the relativistic mass for the rest mass in generalizing a classical formula to a relativistic one. This is not so for the kinetic energy.

3.5 The Equivalence of Mass and Energy

In Section 3-3 we examined an elastic collision, that is, a collision in which the kinetic energy of the bodies remained constant. Now let us consider an inelastic collision. In particular, consider two identical bodies of rest mass m_0, each with kinetic energy K as seen by a particular observer S', which collide and stick together forming a single body of rest mass M_0. The situation before and after the collision in the S'-frame is shown in Fig. 3-6: here, before collision, bodies A and B each have a speed u', with velocities oppositely directed and along the x'-axis; the combined body C, formed by the collision, is at rest in S', as required by conservation of momentum. In another reference frame S, moving with respect to S' with a speed $v(= u')$ to the left along the common x-x' axis, the combined body C will have a velocity of magnitude v directed to the right along x. Body A will be stationary before collision in this frame and body B will have a speed u_B. The situation in the S-frame is shown in Fig. 3-7.

The velocity u_B in the S-frame can be obtained from the relativistic velocity transformation equation, Eq. 2-18, as

$$u_B = \frac{u' + v}{1 + u'v/c^2} = \frac{u' + u'}{1 + u'^2/c^2} = 2u'/(1 + u'^2/c^2).$$

The relativistic mass of B in the S-frame is therefore

$$m_B = \frac{m_0}{\sqrt{1 - u_B^2/c^2}} = \frac{m_0(1 + u'^2/c^2)}{(1 - u'^2/c^2)},$$

as the student can verify by substitution. In S, the combined mass C travels

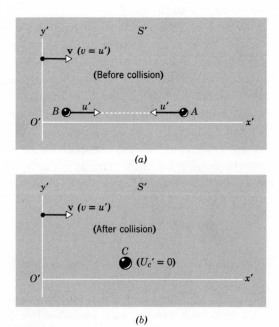

(a)

(b)

Figure 3-6. A particular inelastic collision as viewed by observer S', (*a*) before the collision, and (*b*) after the collision.

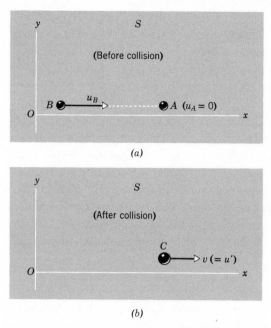

Figure 3-7. The same collision as in Fig. 3-6 as viewed by observer S, (a) before the collision, and (b) after the collision.

at a speed $v(= u')$ after collision, since it was stationary in S'. Hence, applying conservation of relativistic momentum in the x-direction in this frame (the y-component of momentum is automatically conserved), we have

$$\text{(before)} = \text{(after)}$$

$$\frac{m_0}{\sqrt{1 - u_B^2/c^2}} u_B + 0 = \frac{M_0}{\sqrt{1 - v^2/c^2}} v.$$

With $v = u'$ and u_B as given above, this becomes

$$\frac{m_0(1 + u'^2/c^2)}{(1 - u'^2/c^2)} \cdot \frac{2u'}{(1 + u'^2/c^2)} = \frac{M_0 u'}{\sqrt{1 - u'^2/c^2}}$$

whence

$$M_0 = \frac{2m_0}{\sqrt{1 - u'^2/c^2}}. \tag{3-25}$$

The rest mass of the combined body is *not* the sum of the rest masses of the original bodies ($2m_0$) but is *greater* by an amount

$$M_0 - 2m_0 = 2m_0\left(\frac{1}{\sqrt{1 - u'^2/c^2}} - 1\right). \tag{3-26a}$$

Before the collision, the bodies had kinetic energy in S' equal to

$$K_A + K_B = 2K = 2m_0c^2\left(\frac{1}{\sqrt{1 - u'^2/c^2}} - 1\right) \tag{3-26b}$$

but all the kinetic energy disappeared on collision. In its place, after the collision, there appears some form of internal energy, such as heat energy or excitation energy. We now see that this extra internal energy results in the rest mass (inertia) of the combined body being greater than the total rest mass (inertia) of the two separate bodies. Thus, rest mass is equivalent to energy (rest-mass energy) and must be included in applying the conservation of energy principle. This result follows from the Lorentz transformation and the conservation of momentum principle which were used in arriving at it.

From Eqs. 3-26a and 3-26b we see that $K_A + K_B = (M_0 - 2m_0)c^2$, which shows directly, in this case, that the energy associated with the increase in rest mass after the collision, $\Delta m_0 c^2$, equals the kinetic energy present before the collision. We can say, then, that although in an inelastic collision kinetic energy alone is not conserved, *total energy* is conserved. The total energy includes rest-mass energy plus kinetic energy. Furthermore, the conservation of total energy is equivalent to the conservation of relativistic mass. We prove this in the example below, after which we shall draw some important conclusions.

▶ **Example 4.** (*a*) Show that, in both frames S and S', the total energy is conserved in the completely inelastic collision of Figs. 3-6 and 3-7.

Consider first the S'-frame (Fig. 3-6).

Before the collision the total energy is

$$2(m_0 c^2 + K) = 2m_0 c^2 / \sqrt{1 - u'^2/c^2},$$

where we have used Eq. 3-16b on the right.

After the collision the total energy is

$$M_0 c^2 = \left(\frac{2m_0}{\sqrt{1 - u'^2/c^2}} \right) c^2 = 2m_0 c^2 / \sqrt{1 - u'^2/c^2},$$

where we have used Eq. 3-25.

Hence, the total energy is conserved in the collision in frame S'.

Now consider the S-frame (Fig. 3-7).

Before the collision, the total energy is

$$m_0 c^2 + (m_0 c^2 + K_B) = 2m_0 c^2 + m_0 c^2 \left[\frac{1}{\sqrt{1 - u_B^2/c^2}} - 1 \right]$$

$$= 2m_0 c^2 + m_0 c^2 \left[\frac{2u'^2/c^2}{1 - u'^2/c^2} \right] = \frac{2m_0 c^2}{(1 - u'^2/c^2)}.$$

After the collision, the total energy is

$$M_0 c^2 + K_c = \frac{2m_0}{\sqrt{1 - u'^2/c^2}} c^2 + \frac{2m_0}{\sqrt{1 - u'^2/c^2}} c^2 \left[\frac{1}{\sqrt{1 - v^2/c^2}} - 1 \right],$$

which, with $v = u'$, becomes

$$\frac{2m_0}{\sqrt{1 - u'^2/c^2}} c^2 + \frac{2m_0}{\sqrt{1 - u'^2/c^2}} c^2 \left[\frac{1}{\sqrt{1 - u'^2/c^2}} - 1 \right] = \frac{2m_0 c^2}{(1 - u'^2/c^2)}.$$

Hence, the total energy is conserved in the collision in frame S.

(*b*) Show that the relativistic mass is also conserved in each frame.

Consider first the S'-frame (Fig. 3-6).

Before the collision the relativistic mass is

$$\frac{m_0}{\sqrt{1 - u'^2/c^2}} + \frac{m_0}{\sqrt{1 - u'^2/c^2}} = \frac{2m_0}{\sqrt{1 - u'^2/c^2}}.$$

After the collision the relativistic mass is the same as the rest mass, for $U'_C = 0$; that is,

$$M_0/\sqrt{1 - U_C'^2/c^2} = \left(\frac{2m_0}{\sqrt{1 - u'^2/c^2}}\right)/\sqrt{1 - 0} = \frac{2m_0}{\sqrt{1 - u'^2/c^2}}.$$

Hence, the relativistic mass is conserved in the collision in frame S'.
Now consider the S-frame (Fig. 3-7).
Before the collision the relativistic mass is

$$m_0 + \frac{m_0}{\sqrt{1 - u_B^2/c^2}} = m_0 + m_0\frac{(1 + u'^2/c^2)}{(1 - u'^2/c^2)} = \frac{2m_0}{(1 - u'^2/c^2)}$$

After the collision the relativistic mass is

$$M_0/\sqrt{1 - v^2/c^2} = \left(\frac{2m_0}{\sqrt{1 - u'^2/c^2}}\right)/\sqrt{1 - u'^2/c^2} = \frac{2m_0}{(1 - u'^2/c^2)}.$$

Hence, the relativistic mass is conserved in the collision in frame S. ◀

We have seen that the conservation of total energy is equivalent to the conservation of (relativistic) mass. That is, the invariance of energy implies the invariance of (relativistic) mass. Mass and energy are equivalent; they form a single invariant that we can call mass-energy. Simply by multiplying the mass equations above by the universal constant c^2, we obtain numerically the corresponding energy equations. The relation

$$E = mc^2 \tag{3-27}$$

expresses the fact that mass-energy can be expressed in energy units (E) or equivalently in mass units $(m = E/c^2)$. In fact, it has become common practice to refer to masses in terms of electron volts, such as saying that the rest mass of an electron is 0.51 Mev, for convenience in energy calculations.* Likewise, particles of zero rest mass (such as photons, see below) may be assigned an effective mass equivalent to their energy. Indeed, the mass that we associate with various forms of energy really has all the properties that we have given to mass heretofore, properties such as inertia, weight, contribution to the location of the center of mass of a system, and so forth. We shall exhibit some of these properties later in the chapter (see also Ref. 5).

Equation 3-27, $E = mc^2$, is, of course, one of the famous equations of physics. It has been confirmed by numerous practical applications and theoretical consequences. Einstein, who derived the result originally in another context, made the bold hypothesis that it was universally applicable. He

---

*It should be emphasized that mass is not numerically equal to energy, for their units are different. However, they are physically equivalent quantities which correspond to one another. It is somewhat like the correspondence between the height of a mercury column and the air pressure.

considered it to be the most significant consequence of his special theory of relativity.

If we look back now at our single-particle equations (Section 3-4), we see that they are consistent with the conclusions we draw from two-body collisions. There we defined the total energy of a particle as mc^2 and gave it the symbol E. Then we used the relation $E = mc^2$ (below Eq. 3-21) and found that $dm/dt = (1/c^2)(dK/dt)$. This can be expressed also as

$$\frac{dK}{dt} = c^2 \frac{dm}{dt} \tag{3-28}$$

which states that a change in the kinetic energy of a particle causes a proportionate change in its (relativistic) mass. That is, mass and energy are equivalent, their units* differing by a factor c^2.

If the kinetic energy of a body is regarded as a form of external energy, then the rest-mass energy may be regarded as the internal energy of the body. This internal energy consists, in part, of such things as molecular motion, which changes when heat energy is absorbed or given up by the body, or intermolecular potential energy, which changes when chemical reactions (such as dissociation or recombination) take place. Or the internal energy can take the form of atomic potential energy, which can change when an atom absorbs radiation and becomes excited or emits radiation and is deexcited, or nuclear potential energy, which can be changed by nuclear reactions. The largest contribution to the internal energy is, however, the total rest-mass energy contributed by the "fundamental" particles, which is regarded as the primary source of internal energy. This too, may change, as, for example, in electron-positron creation and annihilation. The rest mass (or proper mass) of a body, therefore, is not a constant, in general. Of course, if there are no changes in the internal energy of a body (or if we consider a closed system through which energy is not transferred) then we may regard the rest mass of the body (or of the system) as constant.

This view of the internal energy of a particle as equivalent to rest mass suggests an extension to a collection of particles. We sometimes regard an atom as a particle and assign it a rest mass, for example, although we know that the atom consists of many particles with various forms of internal energy. Likewise, we can assign a rest mass to any collection of particles in relative motion, in a frame in which the center-of-mass is at rest (i.e., in which the resultant momentum is zero). The rest mass of the system as a whole would include the contributions of the internal energy of the system to the inertia.

Returning our attention now to collisions or interactions between bodies, we have seen that regardless of the nature of the collision the total energy is conserved and that the conservation of total energy is equivalent to the conservation of (relativistic) mass. In classical physics we had two separate conservation principles: (1) the conservation of (classical) mass, as in chemical reactions, and (2) the conservation of energy. In relativity, these merge into one conservation principle, that of conservation of mass-energy. The two

---

*A convenient identity (see Problem 24) is $c^2 = (3 \times 10^8 \text{ m/sec})^2 = 931 \text{ Mev/a.m.u.}$

classical laws may be viewed as special cases which would be expected to agree with experiment only if energy transfers into or out of the system are so small compared to the system's rest mass that the corresponding fractional change in rest mass of the system is too small to be measured.

‣ **Example 5.** One atomic mass unit (1 a.m.u.) is equal to 1.66×10^{-27} kg (approximately). The rest mass of the proton (the nucleus of a hydrogen atom) is 1.00731 a.m.u. and that of the neutron (a neutral particle and a constituent of all nuclei except hydrogen) is 1.00867 a.m.u. A deuteron (the nucleus of heavy hydrogen) is known to consist of a neutron and a proton. The rest mass of the deuteron is found to be 2.01360 a.m.u. Hence, the rest mass of the deuteron is *less than* the combined rest masses of neutron and proton by

$$\Delta m_0 = [(1.00731 + 1.00867) - 2.01360] \text{ a.m.u.} = 0.00238 \text{ a.m.u.,}$$

which is equivalent, in energy units, to

$$\Delta m_0 c^2 = (0.00238 \times 1.66 \times 10^{-27} \text{ kg})(3.00 \times 10^8 \text{ m/sec})^2$$
$$= 3.57 \times 10^{-13} \text{ joules} = 2.22 \times 10^6 \text{ ev}$$
$$= 2.22 \text{ Mev.}$$

When a neutron and a proton at rest combine to form a deuteron, this exact amount of energy is given off in the form of electromagnetic (gamma) radiation. If the deuteron is to be broken up into a proton and a neutron, this same amount of energy must be *added* to the deuteron. This energy, 2.22 Mev, is therefore called the *binding energy* of the deuteron.

Notice that

$$\frac{\Delta m_0}{M_0} = \frac{0.00238}{2.01360} = 1.18 \times 10^{-3} = 0.12 \text{ percent.}$$

This fractional rest-mass change is characteristic of the magnitudes that are found in nuclear reactions.

Example 6. The binding energy of a hydrogen atom is 13.58 ev. That is, the energy one must add to a hydrogen atom to break it up into its constituent parts, a proton and an electron, is 13.58 ev. The rest mass of a hydrogen atom, M_0, is 1.00797 a.m.u. The change in rest mass, Δm_0, when a hydrogen atom is ionized is

$$13.58 \text{ ev} = \frac{13.58 \text{ ev}}{931 \times 10^6 \text{ ev/a.m.u.}} = 1.46 \times 10^{-8} \text{ a.m.u.}$$

so that

$$\frac{\Delta m_0}{M_0} = \frac{1.46 \times 10^{-8}}{1.008} = 1.45 \times 10^{-8} = 1.45 \times 10^{-6} \text{ percent.}$$

Such a fractional change in rest mass is actually smaller than the experimental error in measuring the ratio of the masses of proton and electron, so that in practice we could not detect the change. Thus, in chemical reactions, we could not have detected changes in rest mass and the classical principle of conservation of (rest) mass is practically correct. ◂

In a paper [6] entitled "Does the Inertia of a Body Depend upon its Energy Content," Einstein writes:

"If a body gives off the energy E in the form of radiation, its mass diminishes by E/c^2. The fact that the energy withdrawn from the body becomes energy of radiation evidently makes no difference, so that we are led to the more general conclusion that the mass of a body is a measure of its energy content. . . . It is not impossible that with bodies whose energy-content is variable to a high degree (e.g., with radium salts) the theory may be successfully put to the test. If the theory corresponds to the facts, radiation conveys inertia between the emitting and absorbing bodies."

Experiment has abundantly confirmed Einstein's theory.

Today, we call such a pulse of radiation a photon and may regard it as a particle of zero rest mass. The relation $p = E/c$, taken from classical electromagnetism, is consistent with the result of special relativity for particles of "zero rest mass" since, from Eq. 3-19, $E = c\sqrt{p^2 + m_0 c^2}$, we find that $p = E/c$ when $m_0 = 0$. This is consistent also with the fact that photons travel with the speed of light since, from the relation $E = mc^2 = m_0 c^2 / \sqrt{1 - u^2/c^2}$, the energy E would go to zero as $m_0 \to 0$ for $u < c$. In order to keep E finite (neither zero nor infinite) as $m_0 \to 0$, we must let $u \to c$. Strictly speaking, however, the term zero rest mass is a bit misleading because it is impossible to find a reference frame in which photons (or anything that travels at the speed of light) are at rest (see Question 9). However, if m_0 is determined from energy and momentum measurements as $m_0 = \sqrt{(E/c^2)^2 - (p/c)^2}$, then $m_0 = 0$ when (as for a photon*) $p = E/c$.

The result, that a particle of zero rest mass can have a finite energy and momentum and that such particles must move at the speed of light, is also consistent with the meaning we have given to rest mass as internal energy. For if rest mass is internal energy, existing when a body is at rest, then a "body" without mass has no internal energy. Its energy is all external, involving motion through space. Now, if such a body moved at a speed less than c in one reference frame, we could always find another reference frame in which it *is* at rest. But if it moves at a speed c in one reference frame, it will move at this same speed c in all reference frames. It is consistent with the Lorentz transformation then that a body of zero rest mass should move at the speed of light and be nowhere at rest.

▶ **Example 7.** The earth receives radiant energy from the sun at the rate of 1.34×10^3 watts/m². At what rate is the sun losing rest mass due to its radiation? The sun's rest mass is now about 2.0×10^{30} kg.

If we assume that the sun radiates isotropically, then the total solar-radiation rate equals the radiant energy passing per unit time through a sphere having the radius r of the mean earth-sun separation, 1.49×10^{11} m, or

$$[1.34 \times 10^3 \text{ watts/meter}^2][4\pi(1.49 \times 10^{11} \text{ meters})^2] = 3.92 \times 10^{26} \text{ watts.}$$

Since mass/time equals (energy/time)/(energy/mass), we find, from

$$c^2 = E/m = 8.99 \times 10^{16} \text{ joules/kg,}$$

---

*For students who are unfamiliar with the relation $p = E/c$, found in electromagnetism, the argument can be run in reverse. Start with the relativistic relation $E = m_0 c^2 / \sqrt{1 - u^2/c^2}$. This implies that E approaches infinity if $u = c$, unless $m_0 = 0$. Therefore photons, which by definition have $u = c$, must have $m_0 = 0$. Then from $E = c[p^2 + m_0^2 c^2]^{1/2}$ it follows that photons must satisfy the relation $p = E/c$. That this same result is found independently in classical electromagnetism illustrates the consistency between relativity and classical electromagnetism.

that
$$\frac{3.92 \times 10^{26} \text{ joules/sec}}{8.99 \times 10^{16} \text{ joules/kg}} = 4.36 \times 10^9 \text{ kg/sec}$$

is the rate of loss of solar rest mass.

At this rate, the fractional decrease of solar rest mass is

$$\frac{4.36 \times 10^9 \text{ kg/sec} \times 3.14 \times 10^7 \text{ sec/yr}}{2.0 \times 10^{30} \text{ kg}} = 6.8 \times 10^{-14}/\text{yr}.$$

Example 8. We present here an "elementary derivation of the equivalence of mass and energy" attributable to Einstein [7].

Consider a body B at rest in frame S (Fig. 3-8a). Two pulses of radiation, each of energy $E/2$, are incident on B, one pulse moving in the $+x$-direction, the other in the $-x$-direction. These pulses are absorbed by B, whose energy therefore increases by an amount E; from symmetry considerations, B must stay at rest in S. Now consider the same process relative to S', which moves with a constant speed v relative to S in the negative y-direction. Here (Fig. 3-8b) B moves in the positive y'-direction with speed v. The pulses of radiation are here directed upward in part, making an angle α with the x'-axis. The velocity of B remains unchanged in S' after absorption of the radiation since, as we have seen, B stays at rest in S during this process.

Now let us apply the law of conservation of momentum to the process in S'. The momentum of B in S' is Mv (from classical mechanics) along the positive y'-direction. Each pulse of radiation has energy $E/2$ and momentum $E/2c$ (from classical electromagnetism), the y'-components being $(E/2c) \sin \alpha$. Hence, before absorption takes

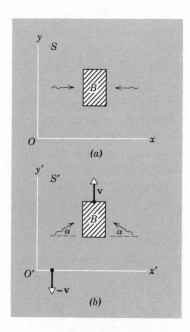

Figure 3-8. (a) Body B is at rest in S. Two pulses of radiation are incident upon it in the directions shown by the arrows. (b) In S', which moves relative to S with speed v in the negative y-direction, body B moves at speed v along $+y'$ and the pulses are directed up at an angle α from the x'-axis, as shown by the arrows.

place, the y'-component of momentum of the system is

$$Mv + 2\left(\frac{E}{2c}\right)\sin\alpha.$$

After absorption, body B has mass M' and its momentum is $M'v$ along the positive y'-direction. Equating y'-components of momentum before absorption to those after absorption we have

$$Mv + \frac{E}{c}\sin\alpha = M'v.$$

Now, since $v \ll c$, $\sin\alpha \cong \alpha = v/c$ (the classical aberration result) so that

$$Mv + \frac{E}{c}\frac{v}{c} = M'v.$$

Hence $$M + \frac{E}{c^2} = M'$$

or $$M' - M = \frac{E}{c^2}.$$

The energy increase E of body B, therefore, is connected to its mass increase, $\Delta M = M' - M$, as

$$E = \Delta Mc^2.$$

In what way (if at all) did Einstein use special relativity theory in this derivation? What approximations were made? ◀

3.6   Relativity and Electromagnetism

In earlier sections we investigated the dynamics of a single particle using the relativistic equation of motion that was found to be in agreement with experiment for the motion of high-speed charged particles. There we introduced the relativistic mass and the total energy, including the rest-mass energy. However, all the formulas we used were applicable in one reference frame, which we called the laboratory frame. Often, as when analyzing nuclear reactions, it is useful to be able to transform these relations to other inertial reference frames, like the center-of-mass frame. In such cases one uses the equations that connect the values of the momentum, energy, mass, and force in one frame S to the corresponding values of these quantities in another frame S', which moves with uniform velocity **v** with respect to S along the common x-x' axes.

Here we simply point out that these equations of transformation can be obtained in a perfectly straightforward way from the transformation equations already derived for the velocity components (Eqs. 2-18 to 2-20). Once we have the force transformations, we can then use the Lorentz force law of classical electromagnetism (Eq. 3-14) to find how the numerical value of electric and magnetic fields depends upon the frame of the observer. The details (see Ref. 8) are of no special interest to us here, but some of the conclusions reached are worth discussing in order to put special relativity theory in proper perspective.

We have seen in the last two chapters how kinematics and dynamics must be generalized from their classical form to meet the requirements of special relativity. And we saw earlier the role that optical experiments played in the development of relativity theory and the new interpretation that is given to such experiments. What remains, therefore, is to investigate classical electricity and magnetism in order to discover what modifications may need to be made there because of relativistic considerations. It turns out that Maxwell's equations are invariant under a Lorentz transformation and do not need to be modified (see Sec. 4.7 of Ref. 8 for proof). This result then completes the original program of finding the transformation (the Lorentz transformation) which keeps the velocity of light constant and finding the invariant form of the laws of mechanics and electromagnetism. The (Einstein) principle of relativity appears to apply to *all* the laws of physics.

Although relativity leaves Maxwell's equations of electromagnetism unaltered, it does give us a new point of view that enhances our understanding of electromagnetism. It is shown clearly in relativity that electric fields and magnetic fields have no separate meaning; that is, **E** and **B** do not exist independently as separate quantities but are interdependent. A field that is purely electric, or purely magnetic, in one inertial frame, for example, will have both electric and magnetic components in another inertial frame. One can find, from the force transformations, just how **E** and **B** transform from one frame to another. These equations of transformation are of much practical benefit for we can solve difficult problems by choosing a reference system in which the answer is relatively easy to find and then transforming the results back to the system we deal with in the laboratory. The techniques of relativity, therefore, are often much simpler than the classical techniques for solving electromagnetic problems.

One striking result we obtain from relativity is this. If all we knew in electromagnetism was Coulomb's law, then, by using special relativity and the invariance of charge, we could prove that magnetic fields must exist. There is no need to postulate magnetic fields separately from electric fields. The magnetic field enters relativity in a most natural way as a field that is produced by a source charge in motion and that exerts a force on a test charge that depends on its velocity relative to the observer. Magnetism is simply a new word, a short-hand designation, for the velocity-dependent part of the force. In fact, starting only with Coulomb's law and the invariance of charge, we can derive* all of electromagnetism from relativity theory—the exact opposite of the historical development of these subjects.

QUESTIONS

1. Can a body be accelerated to the speed of light? Explain.

2. Distinguish between a variable-mass problem in classical physics and the relativistic variation of mass.

3. Can we simply substitute m for m_0 in classical equations to obtain the correct corresponding relativistic equation? Give examples.

---

*See Ref. 9 for a complete development of this approach.

4. Does **F** equal $m\mathbf{a}$ in relativity? Does $m\mathbf{a}$ equal $\dfrac{d}{dt}(m\mathbf{u})$ in relativity?

5. Explain how it happens, in Example 4, that although total energy is conserved in each frame, the value assigned to the total energy in S does not equal numerically the value in S'.

6. We have seen that, in an elastic collision between two spheres, both the rest mass of the sphere and its relativistic mass before collision are equal to the corresponding quantities after collision. What happens to these quantities *during* the collision? (*Hint.* The impulsive forces exerted by the spheres on one another are equivalent to a changing internal elastic potential energy.)

7. How would you expect the relativistic variation of mass to affect the performance of a cyclotron?

8. Is it true that a particle that has energy must also have momentum? What if the particle has no rest mass?

9. If photons have a speed c in one reference frame, can they be found at rest in any other frame? Can photons have a speed other than c?

10. Radiation, being a transfer of energy, involves a transfer of mass and carries momentum. Hence, radiation should exert pressure on bodies it falls upon. Give some examples.

11. Under what circumstances is the rest (or proper) mass of a body a constant quantity? Under what circumstances can we speak of a varying proper mass?

12. A hot metallic sphere cools off on a scale. Does the scale indicate a change in rest mass?

13. What role does potential energy play in the equivalence of mass and energy?

14. A spring is kept compressed by tying its ends together tightly. It is then placed in acid and dissolves. What happens to its stored potential energy?

PROBLEMS

1. Consider a box at rest with sides a, b, and c, as shown in Fig. 3-9. Its rest mass is m_0, and its rest mass per unit volume is $\rho_0 = m_0/abc$. (*a*) What is the volume of the box as viewed by an observer moving relative to the box with speed u in the x-direction? (*b*) What is the mass measured by this observer? (*c*) What is the density of the box, in terms of ρ_0, as measured by this observer? (*d*) Show that when $u \ll c$, the density measured by the observer is the classical result.

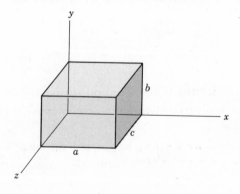

Figure 3-9.

2. What is the speed of an electron whose kinetic energy equals its rest energy? Does the result depend on the mass of the electron?

3. The average lifetime of μ-mesons at rest is 2.3×10^{-6} sec. A laboratory measurement on μ-mesons yields an average lifetime of 6.9×10^{-6} sec. (a) What is the speed of the mesons in the laboratory? (b) The rest mass of a μ-meson is $207\ m_e$. What is the effective mass of such a meson when moving at this speed? (c) What is its kinetic energy? What is its momentum?

4. An electron is accelerated in a synchrotron to a kinetic energy of 1.0 Bev. (a) What is the effective mass of this electron in terms of its rest mass? (b) What is the speed of this electron in terms of the speed of light? (c) What are the answers to (a) and (b) if the kinetic energy of the electron is 1.0 Mev instead?

5. Compute the speed of (a) electrons and (b) protons which fall through an electrostatic potential difference of 10 million volts. (c) What is the ratio of relativistic mass to rest mass in each case?

6. Prove that if $u/c \ll 1$, the kinetic energy K of a moving particle will always be much less than its rest energy $m_0 c^2$.

7. Derive the relation $K = (m - m_0)c^2$ for motion in three dimensions.

8. (a) Verify Eq. 3-18a connecting K and p by eliminating u between Eqs. 3-11 and 3-16b. (b) Derive the following useful relations between p, E, K, and m_0 for relativistic particles, starting from Eqs. 3-18a and 3-18b.

$$(1)\ K = c\sqrt{m_0^2 c^2 + p^2} - m_0 c^2.$$

$$(2)\ p = \frac{\sqrt{K^2 + 2m_0 c^2 K}}{c}.$$

$$(3)\ m_0 = \frac{\sqrt{E^2 - p^2 c^2}}{c^2}.$$

9. (a) What potential difference will accelerate electrons to the speed of light, according to classical physics? (b) With this potential difference, what speed would an electron acquire relativistically? (c) What would its mass be at this speed? Its kinetic energy?

★10. In general (see Eq. 3-22) the acceleration **a** is not parallel to the force **F** in relativity. (a) Show that there are two simple special cases in which the acceleration *is* parallel to the force, namely when **F** is parallel to the velocity **u** and when **F** is perpendicular to the velocity **u**. (b) Give a physical example of each such special case. (c) In view of the fact that we can always resolve a force into two such components (one parallel to **u** and one perpendicular), why is **F** not always parallel to **a**?

11. (a) Show that when $u/c < 1/10$, then $K/m_0 c^2 < 1/200$ and the classical expressions for kinetic energy and momentum, that is, $K = \frac{1}{2}m_0 u^2$ and $p = m_0 u$, may be used with an error of less than 1 percent. (b) Show that when $u/c > 99/100$, then $K/m_0 c^2 > 7$ and the relativistic relation $p = E/c$ for a zero rest-mass particle may be used for a particle of rest mass m_0 with an error of less than 1 percent.

12. (a) Show that a particle which travels at the speed of light must have a zero rest mass. (b) Show that for a particle of zero rest mass, $u = c$, $K = E$, and $p = E/c$.

13. Make a plot of the total energy E versus the momentum p for a particle of rest mass m_0 for the cases (a) classical particle (b) relativistic particle and (c) zero rest-mass particle. (d) In what region does curve (b) approach curve (a) and in what region does curve (b) approach curve (c)? (e) Explain briefly the physical significance of the intercepts of the curves with the axes and of their slopes (derivatives).

14. A 0.50 Mev electron moves at right angles to a magnetic field in a path whose radius of curvature is 2.0 cm. (*a*) What is the magnetic induction *B*? (*b*) By what factor does the effective mass of the electron exceed its rest mass?

15. A cosmic-ray proton of kinetic energy 10 Bev approaches the earth in its equatorial plane in a region where the earth's magnetic induction has a value 5.5×10^{-5} webers/m². What is the radius of its curved path in that region?

★16. (*a*) Show that the angular frequency of a charge moving in a uniform magnetic field is given by $\omega = (qB/m_0) \sqrt{1 - u^2/c^2}$. (*b*) Compare this with the classical result upon which some cyclotron designs are based and explain qualitatively a possible way that the design can be modified relativistically.

17. The general magnetic field in the solar system is 2×10^{-19} tesla (a tesla is one weber/meter²). Find the radius of curvature of a cosmic-ray proton of kinetic energy 10 Bev in such a field. Compare this radius to the radius of the earth's orbit around the sun (1.49×10^{11} m).

18. Ionization measurements show that a particular charged particle is moving with a speed given by $\beta = u/c = 0.71$, and that its radius of curvature is 0.46 m in a field of magnetic induction of 1.0 tesla. Find the mass of the particle and identify it.

19. In a high-energy collision of a primary cosmic-ray particle near the top of the Earth's atmosphere, 120 km above sea level, a π^+ meson is created with a total energy of 1.35×10^5 Mev, travelling vertically downward. In its proper frame it disintegrates 2.0×10^{-8} sec after its creation. At what altitude above sea level does the disintegration occur? The π-meson rest-mass energy is 139.6 Mev.

20. The "effective mass" of a photon (bundle of electromagnetic radiation of zero rest mass and energy $h\nu$) can be determined from the relation $m = E/c^2$. Compute the "effective mass" for a photon of wavelength 5000 Å (visible region), and for a photon of wavelength 1.0 Å (X-ray region).

21. Using data in Example 7, find the pressure of solar radiation at the earth.

22. What is the equivalent in energy units of one gram of a substance?

23. A ton of water is heated from the freezing point to the boiling point. By how much, in kilograms and as a percentage of the original mass, does its mass increase?

24. (*a*) Prove that 1 a.m.u. = 931.5 Mev/c^2. (*b*) Find the energy equivalent to the rest mass of the electron, and to the rest mass of the proton.

25. (*a*) How much energy is released in the explosion of a fission bomb containing 3.0 kg of fissionable material? Assume that 0.1 percent of the rest mass is converted to released energy. (*b*) What mass of TNT would have to explode to provide the same energy release? Assume that each mole of TNT liberates 820,000 calories on exploding. The molecular weight of TNT is 0.227 kg/mole. (*c*) For the same mass of explosive, how much more effective are fission explosions than TNT explosions? That is, compare the fractions of rest mass converted to released energy for the two cases.

26. The nucleus C^{12} consists of six protons (H^1) and six neutrons (n) held in close association by strong nuclear forces. The rest masses are

$$
\begin{array}{ll}
C^{12} & 12.000000 \text{ a.m.u.,} \\
H^1 & 1.007825 \text{ a.m.u.,} \\
n & 1.008665 \text{ a.m.u.}
\end{array}
$$

How much energy would be required to separate a C^{12} nucleus into its constituent protons and neutrons? This energy is called the binding energy of the C^{12} nucleus. (The masses are really those of the neutral atoms, but the extranuclear electrons have relatively negligible binding energy and are of equal number before and after the breakup of C^{12}.)

27. Two identical objects, each of rest mass m_0, moving with equal but opposite velocities of $0.6c$ in the laboratory frame, collide and stick together. The resulting compound particle has a rest mass M_0. Express M_0 in terms of m_0.

28. A body of rest mass m_0, travelling initially at a speed $0.6c$, makes a completely inelastic collision with an identical body initially at rest. (a) What is the rest mass of the resulting single body? (b) What is its speed?

29. Two high-speed particles, each having rest mass m_0, approach each other on course for a head-on collision. One had a speed $0.8c$ and the other a speed $0.6c$. Assuming that the resulting collision is *completely inelastic* answer the following questions in terms of m_0 and c. (a) What is the momentum after collision? (b) What value would Newtonian mechanics predict? (c) What is the value of the total energy after collision? (d) What is the total rest mass after collision? (e) What is the kinetic energy after collision?

30. Consider the following elastic collision: particle A has rest mass m_0 and particle B has rest mass $2m_0$; before the collision, particle A moves in the $+x$-direction with a speed of $0.6c$ and particle B is at rest; after the collision, particle A is found to be moving in the $+y$-direction, and particle B is found to be moving at an angle δ to the $+x$-direction. Write down the three equations (do not solve them) from which we could determine the angle δ and the speeds of A and B after the collision.

31. An excited atom of mass m_0, initially at rest in frame S, emits a photon and recoils. The internal energy of the atom decreases by ΔE and the energy of the photon is $h\nu$. Show that $h\nu = \Delta E \, (1 - \Delta E/2m_0c^2)$.

32. The nucleus of a carbon atom initially at rest in the laboratory goes from one state to another by emitting a photon of energy 4.43 Mev. The atom in its final state has a rest mass of 12.0000 atomic mass units. (1 a.m.u. corresponds to 931.478 Mev). (a) What is the momentum of the carbon atom after the decay, as measured in the laboratory? (b) What is the kinetic energy, *in Mev*, of the carbon atom after the decay, as measured in the laboratory system?

33. A body of mass m at rest breaks up spontaneously into two parts, having rest masses m_1 and m_2 and respective speeds v_1 and v_2. Show that $m > m_1 + m_2$, using conservation of mass-energy.

34. A charged π-meson (rest mass $= 273 \, m_e$) at rest decays into a neutrino (zero rest mass) and a μ-meson (rest mass $= 207 \, m_e$). Find the kinetic energies of the neutrino and the μ-meson.

35. R and E are two experimenters at rest with respect to one another at different points in space. R and E "fire" particles at each other, each particle leaving its gun with a relative speed of $0.6c$. E reports the following observations. For each of these, state the corresponding observations reported by *an observer at rest with respect to R's particles*. (a) The distance from R to E is 10^4 m; (b) The speed of E's particles is $0.6c$; (c) Two particles collided; momentum and kinetic energy are conserved; (d) E's particles are scattered through an angle θ; (e) E fires particles at the rate of 10^4 per second.

REFERENCES

1. A. H. Bucherer, *Ann. Physik,* **28**, 513 (1909).

2. R. Fleischmann and R. Kollath, "Method for Measurement of the Charge of Fast Moving Electrons," *Zeitschrift fur Physik,* **134**, 526 (1953).

3. R. Kollath and D. Menzel, "Measurement of the Charge on Moving Electrons," *Zeitschrift fur Physik,* **134**, 530 (1953).

110

4. W. Bertozzi, "Speed and Kinetic Energy of Relativistic Electrons," *Am. J. Phys.* **32,** 551 (1964).

5. R. T. Weidner, "On Weighing Photons," *Am. J. Phys.* **35,** 443 (1967).

6. *The Principle of Relativity,* Dover Publications, 1923, p. 29. (A collection of original papers by Einstein, Lorentz, Weyl, and Minkowski.)

7. Albert Einstein, *Out of My Later Years,* Philosophical Library, New York, 1950, pp. 116–119.

8. Robert Resnick, *Introduction to Special Relativity,* John Wiley, New York, 1968.

9. Edward M. Purcell, *Electricity and Magnetism,* McGraw-Hill, New York, 1965.

Introduction to Chapters 4 to 7

At a meeting of the German Physical Society in 1900, Max Planck read his paper "On the Theory of the Energy Distribution Law of the Normal Spectrum." This paper, which at first attracted little attention, was the start of a revolution in physics; its date of presentation has come to be called the "birthday of quantum theory." It was not until a quarter of a century later that the modern theory of quantum mechanics, the basis of our present understanding, was developed. Many paths converged on this understanding, each showing another aspect of the breakdown of classical physics. In the next four chapters we examine the major milestones along the way. The phenomena we discuss are all important in their own right and span all the classical disciplines, such as mechanics, thermodynamics, statistical mechanics, and electromagnetism (including light). Their contradiction of classical laws and their resolution on the basis of quantum ideas show us the need for a sweeping new theory and give us a deeper conceptual understanding of the theory that eventually emerged.

CHAPTER 4

Energy Quantization

4.1 Introduction

As is true with relativity theory, the quantum theory represents a generalization of classical physics that includes the classical laws as special cases. Just as relativity extends the range of application of physical laws to the region of high speeds, so quantum mechanics extends the range to the region of small dimensions. And, just as a universal constant of the fundamental significance, c, characterizes relativity, so a universal constant of fundamental significance, h, now called Planck's constant, characterizes quantum physics. It was in the context of thermal radiation that Planck introduced this constant in his 1900 paper. Let us now examine this phenomenon. It will lead us ultimately to an extremely important quantum concept—the discreteness of energy.

4.2 Thermal Radiation

The radiation emitted by a body as a result of its temperature is called *thermal radiation*. All bodies emit such radiation to their surroundings and absorb such radiation from it. If a body is at first hotter than the surroundings it will cool off because its rate of emitting energy exceeds its rate of absorbing energy. When thermal equilibrium is reached, the rates of emission and absorption are equal.

Matter in a condensed state (i.e., solid or liquid) emits a continuous spectrum of radiation. The details of the spectrum depend strongly on the temperature. At ordinary temperatures most bodies are visible to us not by their emitted light but by the light they reflect. If no light shines on them we cannot see them. At very high temperatures, however, bodies are self-luminous. We can see them glow in a darkened room (hot coals, e.g.). But even at temperatures as high as several thousand degrees Kelvin, well over 90% of the emitted thermal radiation is invisible to us, being in the infrared part of the electromagnetic spectrum. Self-luminous bodies are quite hot, therefore.

113

If we were to steadily raise the temperature of a hot body, we would observe two principal effects: (1) the higher the temperature, the more the thermal radiation emitted—at first the body appears dim, then it glows intensely; and (2) the higher the temperature, the higher the frequency of that part of the spectrum radiating most intensely—the predominant color of the hot body shifts from "red heat" to "white heat" to "blue heat." Since the quality of its spectrum depends on the temperature, we can estimate the temperature of a hot body, such as a star or a glowing chunk of steel, by analyzing the radiation it emits. There is a continuous spectrum of radiation emitted, the eye seeing chiefly the color corresponding to the most intense emission in the visible region.

The detailed form of the spectrum of the thermal radiation emitted by a hot body at a given temperature depends somewhat upon the composition of the body. There is one class of hot bodies, however, called *black bodies,* which emit thermal radiation with the same spectrum at a given temperature regardless of the details of their composition. Such bodies have surfaces that absorb all the thermal radiation incident upon them and, because they do not reflect light, appear black. An object coated with a diffuse layer of black pigment, such as lamp black or bismuth black, is (nearly) a black body. A cavity in a body, open to the outside by a small hole, is also a black body, as we shall soon explain. A theoretical understanding of black body, or cavity, radiation was a major goal of physicists before the turn of the present century.

Consider first the experimental observations. The spectral distribution of black body radiation is described by a quantity $\mathcal{R}_T(\nu)$, called the *spectral radiancy,* which is defined so that the quantity $\mathcal{R}_T(\nu)\,d\nu$ is the rate at which energy is radiated per unit area of a surface at absolute temperature T for frequencies in the interval ν to $\nu + d\nu$. In Fig. 4-1, we show the experimentally observed dependence of $\mathcal{R}_T(\nu)$ on ν and T. For a given value of the frequency ν, we see that the spectral radiancy $\mathcal{R}_T(\nu)$ increases with increasing temperature T. If we integrate the quantity $\mathcal{R}_T(\nu)$ over all frequencies ν we obtain the total energy emitted per unit time per unit area from a black body at temperature T. This quantity

$$R_T = \int_0^\infty \mathcal{R}_T(\nu)\,d\nu, \tag{4-1}$$

is called the *radiancy,* appropriate units for it being watts/m^2. It can be interpreted as the area under a curve in Fig. 4-1, from which we see that it increases rapidly as the temperature increases. The exact dependence of radiancy on temperature is given by *Stefan's law,*

$$R_T = \sigma T^4, \tag{4-2}$$

in which

$$\sigma = 5.67 \times 10^{-8} \text{ watt}/(\text{m}^2 {}^\circ\text{K}^4)$$

is a universal constant called the Stefan-Boltzman constant. We also see from Fig. 4-1 that as the temperature T increases, the spectral distribution of frequencies shifts to higher values. If the frequency ν at which $\mathcal{R}_T(\nu)$ reaches

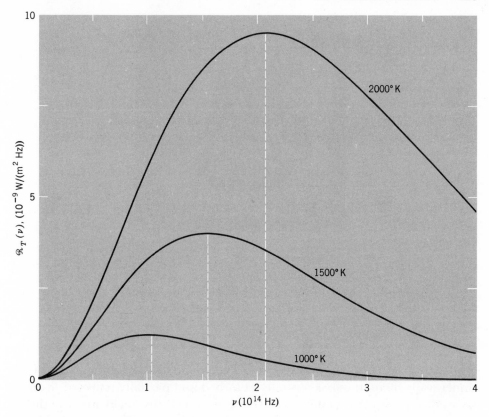

Figure 4-1. The observed spectral radiancy of a black body $\mathcal{R}_T$, as a function of frequency ν of radiation, shown for black body temperatures of 1000°, 1500°, and 2000° Kelvin. The total energy emitted per unit time per unit area (radiancy R_T, or area under the curve) increases rapidly with temperature. Note that the frequency of maximum spectral radiancy (dashed line) increases linearly with temperature (Wien's displacement law). The visible region of the spectrum is off scale to the right, yellow being at about 5×10^{14} Hz. Most of the radiation is in the infrared at these temperatures.

its maximum value is called ν_{max}, then as T increases ν_{max} is displaced toward higher frequencies. This relation,

$$\nu_{max} \propto T, \tag{4-3a}$$

is called *Wien's displacement law*. These experimental results are consistent with the observations we discussed earlier, namely that the quantity of thermal radiation increases rapidly with temperature (a hot body radiates much more heat energy at higher temperatures) and that the principal frequency of the radiation becomes higher with increasing temperature (the color of a hot body changes from red to white to blue).

Most black bodies used in laboratory experiments are cavities (ovens) having a very small opening. Let's explain why such a cavity, shown schematically in Fig. 4-2, is a black body. Radiation outside the cavity can enter it through the hole. It then strikes the inner wall, where it is partially absorbed

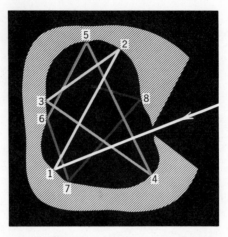

Figure 4-2. A cavity in a body connected by a small hole to the outside. Radiation incident upon the hole is partly absorbed on each reflection and completely absorbed after many successive reflections on the inner surface of the cavity. (Note the diminution in intensity of the successive reflections.) The hole absorbs like a black body. The reverse process, in which radiation leaving the hole is built up of contributions from the inner surface, is discussed in Example 2. The hole emits like a black body.

and partially reflected. The reflected part then strikes another part of the inner wall where it, too, is partially absorbed and partially reflected. Eventually, therefore, after many successive reflections, the incident radiation is totally absorbed by the wall. Because the area cut out by the hole is such a very small part of the total area of the cavity wall, we can safely neglect the (very small) amount of the incoming radiation that can be reflected back out through the hole. For practical purposes we can say that all the radiation incident on the hole from the outside is absorbed by it. Therefore, the hole behaves just as the surface of a black body behaves, i.e., it absorbs all the radiation incident on it. Indeed, at low temperatures the hole appears black. If we raise the temperature by heating the cavity walls uniformly, the hole will become self-luminous. The inner walls emit thermal radiation into the cavity and some very small part of this radiation will emerge from the interior through the hole. Since the hole acts like a black surface, the spectrum of radiation emitted by it will be characteristic of a black body.

Of course, the thermal radiation emitted by the hole is just a specimen of the radiation filling the cavity. Therefore, the radiation inside the cavity also has a spectrum characteristic of a black body. Since the hole acts as a black surface, the spectrum emitted by the hole in the cavity whose walls are at a temperature T can be described, as before, by the spectral radiancy $\mathcal{R}_T(\nu)$. But the spectrum of radiation *inside* the cavity, called *cavity radiation*, is more conveniently described by the energy density $\rho_T(\nu)$, which gives the energy in the frequency interval ν to $\nu + d\nu$ per unit volume of the cavity at temperature T. These quantities (see Question 3) are proportional to one another; that is,

$$\rho_T(\nu) \propto \mathcal{R}_T(\nu). \tag{4-4}$$

Hence, the radiation inside a cavity whose walls are at temperature T has the same character as the radiation emitted by the surface of a black body at temperature T. It is convenient experimentally to produce a black body spectrum by means of a cavity in a heated body with a hole to the outside; and it is convenient in theoretical work to study black body radiation by analyzing the cavity radiation. In the next section we carry out that analysis.

◆ **Example 1.** (*a*) Since $\lambda \nu = c$, a constant, Wien's displacement law, Eq. 4-3*a* can also be put in the form

$$\lambda_{max} T = \text{constant} \tag{4-3b}$$

where λ_{max} is the wavelength at which the spectral radiancy has its maximum value for a particular temperature T. The experimentally determined value of Wien's constant is 2.898×10^{-3} m°K. If we assume that the stellar surfaces behave like black bodies, we can get a good estimate of their temperature by measuring λ_{max}. For the sun $\lambda_{max} = 5100$ Å, whereas for the North Star $\lambda_{max} = 3500$ Å. Find the surface temperature of these stars.

For the sun, $T = (\text{Wien's constant})/\lambda_{max} = 2.898 \times 10^{-3}$ m°K$/5100 \times 10^{-10}$ m $= 5700$°K.

For the North Star, $T = (\text{Wien's constant})/\lambda_{max} = 2.898 \times 10^{-3}$ m°K$/3500 \times 10^{-10}$ m $= 8300$°K.

At 5700°K the sun's surface is near the temperature at which the greatest part of its radiation lies within the visible region of the spectrum. This suggests that over ages of human evolution our eyes adapted to the sun to become most sensitive to those wavelengths that it radiates most intensely.

(*b*) Using Stefan's law, Eq. 4-2, and the temperatures obtained above, determine the power radiated from 1 cm² of stellar surface.

For the sun,

$$R_T = \sigma T^4 = [5.67 \times 10^{-8} \text{ watt/m}^2°\text{K}^4](5700°\text{K})^4$$
$$= 6000 \text{ watts/cm}^2$$

For the North Star,

$$R_T = \sigma T^4 = [5.67 \times 10^{-8} \text{ watt/m}^2°\text{K}^4](8300°\text{K})^4$$
$$= 27{,}000 \text{ watts/cm}^2.$$

Example 2. The brightness of a surface is a quantity that is proportional to its spectral radiancy. The hole of a cavity whose walls have been heated to a uniform temperature will appear brighter than the outer wall of the cavity when we observe the body by its own radiation (no external radiation incident). Let b_0 represent the brightness of the hole and b that of the wall and prove this statement by obtaining the relation between b_0 and b. Use Fig. 4-2 and assume for simplicity that the wall is polished so that the surface characteristics are the same everywhere.

The cavity radiation is independent of the material of the walls. The radiation emerging from the hole along the direction of the line in Fig. 4-2 that passes through the opening (the arrow would point *outward* in this case) consists of the radiation emitted directly by the inner wall at region 1 plus the radiation reflected out off of region 1 that originated at other areas of the inner surface. Let r represent the fraction of radiation incident on the wall material that would be reflected by it. Then, the brightness of the hole, b_0, viewed through the opening in the direction shown in Fig.

4-2 would be built up of b, the brightness of the wall at 1; of rb, the brightness of the wall at 2 once reflected; of r^2b, the brightness of the wall at 3 twice reflected; etc., so that

$$b_0 = b + rb + r^2b + r^3b + \cdots = b(1 + r + r^2 + r^3 + \cdots).$$

But $(1 + r + r^2 + \cdots) = 1/(1 - r)$, as the student can show simply by longhand division, so that $b_0 = b/(1 - r)$ or

$$b = b_0(1 - r).$$

Here, b_0 is the brightness of the hole and b is that of the wall.

Hence, if r were 0.2, then $b = (1 - r)b_0 = 0.8b_0$ so that the outer wall adjacent to the hole would appear only eight-tenths as bright as the hole itself. And if r were 0.8, the outer wall would appear only two-tenths as bright as the hole itself when observed by its own radiation. If $r = 0$, however, the wall material itself acts as a black body, for then all radiation incident on it would be absorbed (i.e., none would be reflected with $r = 0$). From the above equation, in fact, we obtain $b = b_0$ when $r = 0$.

Clearly, a black body is brighter than any nonblack body when bodies are viewed by their own temperature radiation (see Fig. 4-3). A black body (or cavity hole) is therefore the most efficient emitter of radiation, as well as the most efficient absorber.

Can you reverse the argument above to show that the hole in the cavity absorbs like a black body? ◀

4.3 The Classical Theory and the Planck Theory of Cavity Radiation

Early in 1900, Lord Rayleigh and Sir James Jeans presented a classical calculation of the energy density of cavity (or black body) radiation. The details of their calculation need not concern us here. But an outline of their general procedure will help point out the serious conflict between the classical approach and the experimental results.

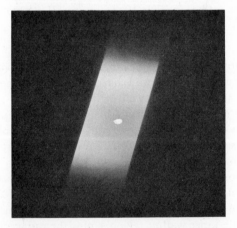

Figure 4-3. Photograph of an incandescent tungsten tube with a small hole drilled in its wall. The radiation emerging from the hole is cavity radiation and is brighter than the surface.

Suppose that we uniformly heat to a temperature T a piece of metal containing a cavity. The electrons in the metallic walls are thermally agitated and emit electromagnetic radiation into the cavity. Thermal equilibrium is established and maintained in the cavity by the absorption and reradiation of energy by the walls. Rayleigh and Jeans showed that the radiation inside such a cavity of volume V consists of standing waves with nodes at the walls. They computed the number of standing waves in the frequency interval ν to $\nu + d\nu$ to be

$$N(\nu)\,d\nu = \frac{8\pi V}{c^2}\,\nu^2\,d\nu \qquad (4\text{-}5)$$

in which c is the velocity of electromagnetic waves. Now, each such standing wave contains energy. The average energy per wave, when the system is in thermal equilibrium, can be determined from the classical law of equipartition of energy (see *Physics* Part I, Sec. 23-8). This states that the average energy is the same for each standing wave in the cavity, independent of its frequency, i.e., the energy is partitioned equally over all frequencies; the value of the average energy $\bar{\varepsilon}$ depends only on the temperature T and is given by

$$\bar{\varepsilon} = kT \qquad (4\text{-}6)$$

where k $(= 1.37 \times 10^{-23}$ joule $°K)$ is the Boltzman constant. To get the average energy content per unit volume of cavity in the frequency interval ν to $\nu + d\nu$, we simply multiply the number of standing waves in the frequency interval by the average energy of a wave and divide by the volume of the cavity. In this way Rayleigh and Jeans found the energy density $\rho_T(\nu)$ to be

$$\rho_T(\nu)\,d\nu = \frac{8\pi\nu^2}{c^3}kT\,d\nu, \qquad (4\text{-}7)$$

which is called *the Rayleigh-Jeans formula for black-body radiation.*

We compare the Rayleigh-Jeans formula, in Fig. 4-4, with the experimental result for a cavity radiator at 1500°K. As the frequency is reduced to zero, the spectrum predicted by the Rayleigh-Jeans classical formula does come closer and closer to the experimentally observed spectrum. However, as the frequency is increased to large values (ultraviolet region of the spectrum), the classical theoretical result diverges enormously from experiment. Indeed, the classical formula predicts an infinite energy density whereas experiment shows that the energy density goes to zero at very high frequencies. This completely erroneous prediction of classical physics was regarded as such a serious shortcoming that it came to be called "the ultraviolet catastrophe" (see also Prob. 10).

Max Planck was able to deduce a theoretical result that is in complete agreement with experiment. Again we can ignore the details for our purpose. Instead, we focus on the central contribution Planck made to modify the classical view. Planck conjectured that the law of equipartition of energy, on which the classical theory is based, might break down in the case of cavity radiation. The classical result agrees with experiment as the frequency goes to zero, so that the prediction of the equipartition law that the average energy

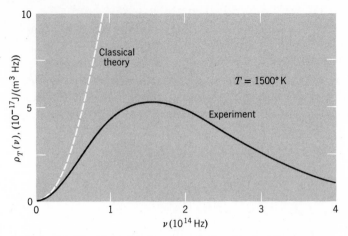

Figure 4-4. The Rayleigh-Jeans prediction (dashed line) compared with the experimental result (solid line) for the energy density of a black body, showing the 'ultraviolet catastrophe'.

of a standing wave $\bar{\varepsilon}$ equals kT can be assumed to be correct in the limit as the frequency goes to zero. On the other hand, in the limit as the frequency goes to infinity the average energy must approach zero, if the classical result is to match experiment there. Instead of assigning a value to the average energy $\bar{\varepsilon}$ that is independent of frequency, as required by the classical law of equipartition of energy (Eq. 4-6), Planck assumed that, for cavity radiation at least, the average energy of the standing waves $\bar{\varepsilon}$ is *dependent* on the frequency.

In the classical calculation of the average energy $\bar{\varepsilon}$ of a large number of things of the same kind in thermal equilibrium with each other at temperature T (a calculation that leads to the equipartition law), it is assumed that the energy ε is a continuous variable. Planck found that if he assumed instead that the energy ε were a *discrete* (noncontinuous) variable then he could obtain the desired behavior for the average energy $\bar{\varepsilon}$. In particular, Planck assumed that the energy could take on only certain discrete values, rather than any value, and that the discrete set of allowed values of the energy was

$$\varepsilon = 0, \, \Delta\varepsilon, \, 2\,\Delta\varepsilon, \, 3\,\Delta\varepsilon, \, 4\,\Delta\varepsilon, \ldots$$

or
$$\varepsilon_n = n\,\Delta\varepsilon \qquad n = 0, 1, 2, \ldots . \tag{4-8}$$

Here $\Delta\varepsilon$ is an assumed uniform interval between successive allowed values of the energy and n takes on only integral values. With this assumption, Planck then calculated that the average energy $\bar{\varepsilon}$ goes to zero if $\Delta\varepsilon$ is chosen to be large and that it goes to kT if $\Delta\varepsilon$ is chosen to be small. That is, large differences in adjacent energies corresponded to the high frequency behavior of cavity radiation of $\bar{\varepsilon}$ and small differences to the low frequency behavior. The simplest assumption to make then is that this difference $\Delta\varepsilon$ is directly proportional to the frequency ν, or

$$\Delta\varepsilon = h\nu \tag{4-9}$$

where h is a constant of proportionality. Hence, from Eq. 4-8, the allowed values of the energy itself are

$$\varepsilon_n = n\, \Delta\varepsilon = nh\nu. \qquad (4\text{-}10)$$

Planck then determined a value of this constant h that would give the best agreement between his theory and the experimental data. He obtained a value nearly the same as the modern value of

$$h = 6.63 \times 10^{-34} \text{ joule-sec}$$
$$= 4.14 \times 10^{-15} \text{ ev-sec.}$$

Ever since, this fundamental constant has been called *Planck's constant*.

The actual formula for $\bar{\varepsilon}$ that Planck arrived at was

$$\bar{\varepsilon}(\nu) = \frac{h\nu}{e^{h\nu/kT} - 1} \qquad (4\text{-}11)$$

rather than the classical result

$$\bar{\varepsilon} = kT. \qquad (4\text{-}6)$$

We compare these in Fig. 4-5; note that $\bar{\varepsilon}(\nu)$ drops from kT to zero in a continuous way with increasing frequency, at first rapidly and then slowly.

When Planck used his result (Eq. 4-11) for $\bar{\varepsilon}$ rather than the classical value $\bar{\varepsilon} = kT$ (Eq. 4-6) in the calculation of the energy density in the cavity radiation spectrum, he found

$$\rho_T(\nu)\, d\nu = \frac{8\pi\nu^2}{c^3} \frac{h\nu}{e^{h\nu/kT} - 1}\, d\nu. \qquad (4\text{-}12)$$

This is the Planck formula for black body radiation. Figure 4-6 shows a comparison of this result of Planck's theory with experiment for a temperature $T = 1595°\text{K}$. The experimental results are in complete agreement with Planck's formula at all temperatures.

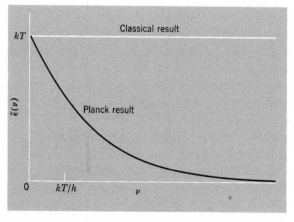

Figure 4-5. Planck's formula for the average value of the energy, $\bar{\varepsilon}$, as a function of frequency, compared with the classical result.

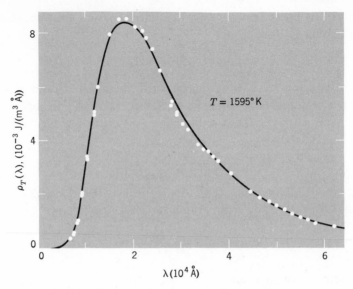

Figure 4-6. Planck's energy density prediction (solid line) compared to the experimental results (circles) of the energy density of a black body. The data were reported by Coblentz in 1916 and apply to a temperature of 1595°K. The author remarked in his paper that after drawing the spectral energy curves resulting from his measurements, "owing to eye fatigue, it was impossible for months thereafter to give attention to the reduction of the data." The data, when finally reduced, led to a value of Planck's constant of 6.57×10^{-34} J-sec.

One can understand the results of Planck's assumption physically in this way. The electromagnetic waves in the cavity originate from radiation given off by electrons that are thermally agitated and oscillate in the walls of the cavity. The electronic oscillators in the cavity wall are pictured classically as radiating their energy smoothly as their motion gradually subsides. Planck assumed instead that an oscillator ejects its radiation in spurts. Thus the energy of an oscillator does not subside continuously but discretely. The allowed energy values of an oscillator must then be discrete and, as it exchanges energy with the cavity radiation, the oscillator emits or absorbs radiant energy only in discrete amounts. Since the discrete energies that an oscillator can emit or absorb are directly proportional to its frequency, the oscillators of low frequency can absorb or emit energy in small packets whereas those of high frequency absorb or emit only large energy packets. Now imagine that the cavity wall is at a low temperature. Then there is sufficient thermal energy to excite the low frequency oscillators, but not the high frequency ones. The high frequency oscillators need to receive much more energy to begin radiating and fewer of them proportionally are activated compared to the low frequency oscillators (the energy is *not* partitioned equally over all frequencies). Hence the walls radiate principally in the long wavelength region and hardly at all in the ultraviolet. As the temperature of the wall is raised there is sufficient thermal energy to activate a larger number of high frequency oscillators and the resulting radiation shifts its character toward higher frequencies, i.e., toward the ultraviolet. Hence the Planck assumptions quite naturally lead to the experimental observations discussed earlier and avoid the ultraviolet catastrophe of the classical analysis.

▶ **Example 3.** (*a*) Show that the Planck radiation law, Eq. 4-12, reduces to the Rayleigh-Jeans formula, Eq. 4-7, for low values of the frequency.

If $h\nu \ll kT$, we can expand the exponential function in the denominator of

$$\rho_T(\nu) = \frac{8\pi h\nu^3}{c^3} \frac{1}{e^{h\nu/kT} - 1}$$

in increasing powers of $h\nu/kT$ and ignore terms beyond the first power, obtaining

$$\rho_T(\nu) = \frac{8\pi h\nu^3}{c^3} \frac{1}{(1 + h\nu/kT + \cdots) - 1} \simeq \frac{8\pi\nu^2}{c^3} kT$$

which is the Rayleigh-Jeans formula. Planck's formula for long-wavelength radiation therefore reduces to the classical formula.

(*b*) What is the high frequency limit of Planck's formula?

In this case, $h\nu/kT \gg 1$. The exponential term is very large compared to one, so that the 1 in the denominator can be ignored. We obtain, therefore,

$$\rho_T(\nu) \simeq \frac{8\pi h\nu^3}{c^3} e^{-h\nu/kT}.$$

This relation, the short-wavelength limit of Planck's law is exactly the form of a relation obtained by Wien much earlier as a fit to experiment in the high frequency end of the spectrum. Wien's relation, obtained by making certain assumptions about the process of radiation from molecules, was

$$\rho_T(\nu) = c_1\nu^3\, e^{-\frac{c_2\nu}{T}}$$

where c_1 and c_2 were experimentally determined constants. Wien's "fit" failed badly at low frequencies (in the far infrared and beyond) and the Rayleigh-Jeans theory failed "catastrophically" at high frequencies (ultraviolet and beyond). Only Planck was able to derive a result that agreed with experiment everywhere.

Example 4. It is convenient in experimental work to express the Planck radiation law in terms of wavelength λ rather than of frequency ν. Obtain $\rho_T(\lambda)$, the wavelength form of Planck's radiation law, from $\rho_T(\nu)$, the frequency form of Planck's law.

The quantity $\rho_T(\lambda)$ is defined from the equality $\rho_T(\lambda)\,d\lambda = -\rho_T(\nu)\,d\nu$. The minus sign indicates that, although $\rho_T(\lambda)$ and $\rho_T(\nu)$ are both positive, $d\lambda$ and $d\nu$ have opposite signs (an increase in frequency gives rise to a corresponding decrease in wavelength).

From the relation $\nu = c/\lambda$ we have $d\nu = (c/\lambda^2)\,d\lambda$, or $d\nu/d\lambda = -(c/\lambda^2)$, so that

$$\rho_T(\lambda) = -\rho_T(\nu)\frac{d\nu}{d\lambda} = \rho_T(\nu)\frac{c}{\lambda^2}.$$

If now we set $\nu = c/\lambda$ in Eq. 4-14 for $\rho_T(\nu)$, we obtain

$$\rho_T(\lambda)\,d\lambda = \frac{8\pi hc}{\lambda^5} \frac{d\lambda}{e^{hc/\lambda kT} - 1}. \tag{4-13}$$

In Fig. 4-7 we show $\rho_T(\lambda)$ versus λ for several different temperatures. The trend from "red heat" to "white heat" to "blue heat" radiation with rising temperatures becomes

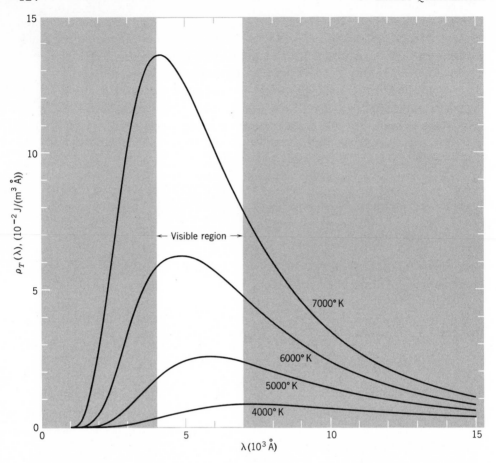

Figure 4-7. Planck's energy density of black body radiation at various temperatures as a function of wavelength. Note that the wavelength at which the curve is a maximum decreases as the temperature increases. The temperatures chosen here are higher than those in the corresponding frequency curves of Fig. 4-1, so that the visible region makes a sizable contribution to the total energy emitted.

clear as the distribution of radiant energy with wavelength is studied for increasing temperatures. ◀

It is worth noting that both Stefan's law (Eq. 4-2) and Wien's displacement law (Eq. 4-3) can be derived from the Planck formula (see Probs. 13 and 14) and, from the experimental results, it is possible to determine values of the constants h and k. Indeed, this was done by Planck, his values agreeing very well with those obtained subsequently by other methods.

4.4 Energy Quantization

In our deduction of the Planck formula, we treated the energy of the standing electromagnetic waves in the cavity as a discrete instead of a continuous quantity. We are already familiar with classical systems, such as sound waves in air columns, in which the frequency of standing waves takes on only discrete values. What is new here is that for any given frequency the *energy* of the wave takes on only discrete values. This quantization of energy does *not* occur in classical theories. Indeed, the classical result that the energy can vary continuously can be regarded as a special case of Planck's result. If in $\Delta\varepsilon = h\nu$ one lets h go to zero, for any frequency ν, then $\Delta\varepsilon$ goes to zero; that is, $\underset{h \to 0}{\Delta\varepsilon \to 0}$ so that energy intervals are not finite and the energy varies continuously when $h = 0$. It is the finite, nonzero value of Planck's constant h, therefore, that distinguishes Planck's result from classical results.

Planck postulated that simple harmonic oscillators can have only total energies ε which take on discrete values given by

$$\varepsilon_n = nh\nu \qquad n = 0, 1, 2, 3, \ldots$$

in which h is a universal constant and ν is the frequency of the oscillation. We can compare the behavior of a Planck oscillator with that of a classical one on an energy level diagram, as shown in Fig. 4-8. In such a diagram we represent each possible energy state by a horizontal line. We choose the distance of such a line from the one of zero energy to be proportional to the energy of the state. According to classical physics an oscillator may have any one of a continuous range of energy values, from zero to infinity, so that the classical energy level diagram of an oscillator is simply a continuum of lines. According to Planck, however, a simple harmonic oscillator can have

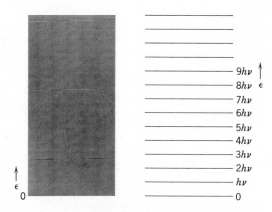

Figure 4-8. *Left:* The allowed energies in a system, oscillating with a frequency ν, are continuous according to classical theory. *Right:* The allowed energies according to Planck's postulate are discrete and equal to $nh\nu$. We say that the energy is quantized, n being the quantum number of an allowed energy state.

The two pictures are compatible (see Example 5), the right picture being merely a clearly resolved view of an extremely narrow energy range of the left picture, greatly magnified.

any one of only a discrete set of energy values, $\varepsilon = 0, h\nu, 2h\nu, 3h\nu, \ldots$, from zero to infinity, so that the Planck energy level diagram consists of a discrete set of lines. In today's terminology, we say that the Planck oscillator is quantized and call each possible state of energy a quantum state. The integer n, which specifies the energy of the state, is called the quantum number of the state.

Ordinary spring-mass systems and physical pendula oscillating with small amplitudes are common examples of simple harmonic oscillators. They do not seem to have discrete energies, but instead appear to pass through a continuous range of energies as their motion dies out. These observations do not necessarily contradict Planck's postulate, however, as we illustrate in the following example.

▶ **Example 5.** Consider a spring-mass system in which a mass of 0.30 kg is suspended from a spring of force-constant $k = 3.0$ nt/m. The system is set oscillating with an amplitude A of 0.10 meter. Gradually the oscillation dies out because of viscous or frictional effects and the total energy of the system is dissipated.

(*a*) Is the energy decrease observed to be continuous or discontinuous?

The frequency of oscillation of the spring-mass system (see *Physics*, Part I, Sec. 15-4) is

$$\nu = \frac{1}{2\pi}\sqrt{\frac{k}{m}} = \frac{1}{2\pi}\sqrt{\frac{3.0 \text{ nt/m}}{0.30 \text{ kg}}} = 0.5 \text{ sec}^{-1}.$$

The total energy of the system (kinetic plus potential) initially is

$$E = \tfrac{1}{2}kA^2 = \tfrac{1}{2}\left(3.0\frac{\text{nt}}{\text{m}}\right)(0.10 \text{ m})^2 = 1.5 \times 10^{-2} \text{ joule}.$$

Now assume that the energy of the pendulum is quantized so that as energy is dissipated the changes in energy take place in discontinuous jumps of magnitude $\Delta E = h\nu$. Then we find that energy losses occur in jumps of

$$\Delta E = h\nu = 6.63 \times 10^{-34} \text{ joule-sec} \times 0.50 \text{ sec}^{-1}$$
$$= 3.3 \times 10^{-34} \text{ joules}$$

whereas the total energy itself is

$$E = 1.5 \times 10^{-2} \text{ joules}.$$

Therefore,

$$\frac{\Delta E}{E} \simeq 2 \times 10^{-32}.$$

Hence, to measure the discreteness in the energy decrease we would need to measure the energy to better than 2 parts in 10^{32}. Clearly, even the most discriminating equipment cannot resolve the energy this finely.

(*b*) Calculate the value of n, the quantum number of the energy state, for the initial energy of the macroscopic oscillator.

From the Planck relation (Eq. 4-10) $\varepsilon_n = n \Delta\varepsilon$, we have

$$n = \frac{E}{\Delta E} = \frac{1.5 \times 10^{-2} \text{ joule}}{3.3 \times 10^{-34} \text{ joule}} = 45 \times 10^{30}.$$

This is an enormous number. If, in Fig. 4-8, we plotted all the horizontal lines of energy from $n = 0$ to $n = 45 \times 10^{30}$ and then compressed the scale to fit the lines into the vertical distance shown in that figure, we would see a continuum of lines.

Hence, the classical continuum can be regarded as a set of discrete lines that are too close for us to resolve experimentally; likewise the quantum discreteness can be regarded as a greatly magnified and clearly resolved view of an extremely narrow energy range of the classical continuum. ◀

This example shows that for ordinary large-scale oscillators, the quantized nature of the energy of the oscillators will not be apparent to us. This result is similar to our not being aware in large-scale experiments of the discrete nature of mass and the quantized nature of charge, i.e., of the existence of atoms and electrons. Macroscopic systems will not reveal whether Planck's postulate is valid or not. The smallness of h makes the graininess in the energy too fine to be distinguished from an energy continuum. Indeed, h might as well be zero for classical systems and, in fact, one way to reduce quantum formulas to their classical limits would be to let $h \to 0$ in these formulas. The fact that Planck's postulate gives the correct cavity radiation formula, however, suggests that for microscopic oscillators (electrons, atoms, etc.) ΔE and E can be of comparable order, so that the quantized nature of energy will reveal itself at the microscopic level. In the next section we examine another system at that level, the atom.

4.5 The Franck-Hertz Experiment

Planck, in his initial work (see Ref. 1), quantized the energy of only the electrons in the walls of the cavity. He did not at first accept the idea that the oscillating electromagnetic waves in the cavity were also quantized. The early Planck theory suggested the idea that, in the process of emission and absorption of radiation, the atoms in the cavity wall behave as though they had quantized energy states. The atom was known at the time to have internal structure; for example, it was known to contain electrons which sometimes could be knocked out of the atom. It was not yet clear, however, just what the internal structure of the atom was, or even what all its parts were. But whether one regarded the atom as a kind of oscillator or not, it was clear that it could absorb and emit energy.

When two atoms collide, as in a gas, it could happen then that some of their translational kinetic energy is converted into internal energy of one or both of the atoms. Should this happen we call the collision inelastic, for translational kinetic energy is not conserved in the collision. If the internal energy of an atom cannot have any of a continuous set of values, however, but only certain discrete ones, then an inelastic collision could take place only if enough kinetic energy were available to raise one of the colliding atoms from its original internal energy state to the next higher allowed internal energy state. Otherwise, the collision would be elastic, none of the kinetic energy being absorbed internally. If an atom is raised from its lowest internal energy state to some higher one, however, the atom can thereafter give up this absorbed energy by emitting radiation whose energy equals the difference in energy between the upper and the lower internal energy states.

Evidence that the internal energy of an atom *is* quantized already existed in experiments with gas collisions, though it was not generally recognized at the time. In classical kinetic theory, for example, atoms are regarded as having no internal structure and as colliding elastically with one another. At ordinary temperatures this assumption is justified, for the experimental results agree with the kinetic theory predictions. Recall (see *Physics*, Part 1, Sec. 23-5), that in a gas the average translational kinetic energy of an atom is $\frac{3}{2}kT$. At room temperature this corresponds to an energy of about $\frac{1}{25}$ ev (see Prob. 15). If the internal energy states of the atom are separated by energies much greater than this, then inelastic collisions cannot occur. Only at temperatures high enough to give the atoms an average translational kinetic energy comparable to the energy difference between the lowest and the first allowed excited state of the atom will the internal structure of the atom change and the collisions become inelastic. Then an appreciable number of the atoms can absorb enough energy through inelastic collisions to be raised to an excited state. We can detect this because, after an interval, radiation corresponding to the absorbed energy will be emitted. The effect was, in fact, observed. For most gases the required temperature is of the order of 10^4 °K, corresponding to excitation energies of several electron volts (see Prob. 15). Hence, in retrospect, one may say that early evidence that the internal energy of an atom is quantized existed in experiments with gas collisions and that the seeds of energy quantization lay in the kinetic theory of gases (see Ref. 3).

In 1914, James Franck and Gustav Hertz obtained direct evidence that the internal energy states of atoms are quantized. Their experiment showed that electrons lost energy only in discrete amounts in collisions with atoms. The Franck-Hertz apparatus, shown schematically in Fig. 4-9, consisted of a tube, filled with mercury vapor, containing a heated filament cathode, a plate anode, and an accelerating grid near the plate. The filament temperature is set so that thermally excited electrons are emitted from its surface with little kinetic energy; the grid voltage is set to accelerate the electrons toward the plate. A small retarding potential is put between plate and grid so that electrons that pass through openings in the grid with very small kinetic energy will not reach the plate. Then the experimenter measures the variation

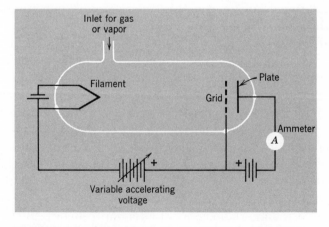

Figure 4-9. Schematic of apparatus used by Franck and Hertz.

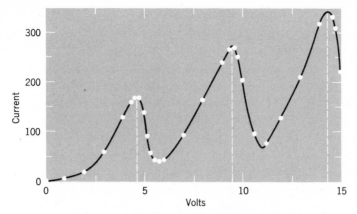

Figure 4-10. The current, measured in the Franck-Hertz experiment, as a function of accelerating voltage. The break at 4.9 volts corresponds to electrons exciting the atoms just in front of the grid. The breaks at 9.8 volts and 14.7 volts correspond to electrons exciting the first energy state of the atoms on several separate occasions in their trip from filament to plate.

in the electron beam current through the plate, by means of an ammeter in the plate circuit, as the accelerating voltage V is increased.

The results obtained with the tube containing mercury vapor, shown in Fig. 4-10, were at first similar to what one expects in an evacuated tube; that is, when the accelerating voltage is low the current increased as the voltage V was increased. However, when V reached the value 4.9 volts, the current dropped sharply. The interpretation is that electrons lose most of their kinetic energy in inelastic collisions with the mercury atoms in the tube once they reach the kinetic energy of 4.9 ev just in front of the grid; they then have little energy left and the retarding potential prevents them from reaching the plate—hence the drop in plate current. The Hg atoms absorb the energy and are said to be in an excited state. As the voltage increases above 4.9 volts, the electrons have enough kinetic energy after the excitation process to overcome the retarding potential and to reach the plate, so the current rises again. When the accelerating voltage was made large enough so that electrons had enough kinetic energy to make *two* separate inelastic collisions with Hg atoms in their trip from cathode to anode, the current fell again. Hence, as the voltage was increased, Franck and Hertz observed a series of drops in current spaced at about equal intervals of 4.9 ev.

All this suggests that the first excited energy state of mercury is about 4.9 ev above its lowest energy state and that Hg atoms are not able to absorb energy from the incident electrons unless they have at least an energy of 4.9 ev. The Hg atoms that are excited would be expected to give up this energy by emitting radiation as they make a transition from the first excited state back to the lowest (ground) state. In later experiments Hertz observed the spectrum emitted by mercury vapor in the tube. He found that when the kinetic energy of the incident electrons was less than 4.9 ev no spectral lines were emitted at all. However, as the 4.9 ev energy was reached a single spectral line of Hg appeared in the spectrum. Even when the electron kinetic energy was somewhat greater than 4.9 ev only this one line, the 2536 Å line, was seen in the spectrum. This is exactly the behavior expected if the internal energy of the mercury atom is quantized.

With grid arrangements different than the one shown in Fig. 4-9, the electrons can be made to gain energy quickly enough so that they might not collide with a Hg atom until their energy is able to excite still higher energy states than the first. Indeed, Hertz found with such an arrangement that other lines appeared in the spectrum just as the accelerating voltage was made large enough to excite the state believed to give rise to that line in transitions to a lower state. Therefore, the Franck-Hertz experiments gave striking evidence for the quantization of the internal energy of atoms. In fact, by measuring the breaks in the current versus voltage curve, the energy differences between the allowed quantum states can be measured directly in a Franck-Hertz experiment. It is found (see Prob. 19) that the allowed energy states of the mercury atom are not spaced equally, as was the case for the harmonic oscillator. Although its energy is quantized, the mercury atom is not as simple a system as the oscillator. We shall see in Chapter 7 that as the Franck-Hertz experiments were being done, Neils Bohr was independently advancing his ideas of atomic structure to construct a theory of the atom that predicted quantization of the energy.

Hence, from Planck's analysis of cavity radiation and from the Franck-Hertz experiments, we are faced with a conclusion that contradicts classical theory. *In some systems energy cannot be changed in a continuous way or by an arbitrary amount; instead the energy of the system can take on only certain discrete values. When such a system exchanges energy with its environment it does so in discrete amounts and in a discontinuous way.*

4.6 A Bit of Quantum History

We have mentioned that in Planck's original analysis he quantized the energy of the electrons in the cavity walls and only later accepted the quantization of the cavity radiation itself. Indeed, at first Planck was unsure whether his introduction of the constant *h* was only a mathematical device or a matter of deep physical significance. Much later, in a letter to R. W. Wood, Planck called his limited postulate "an act of desperation." "I knew," he wrote, "that the problem (of the equilibrium of matter and radiation) is of fundamental significance for physics; I knew the formula that reproduces the energy distribution in the normal spectrum; a theoretical interpretation *had* to be found at any cost, no matter how high." For more than a decade Planck tried to fit the quantum idea into classical theory. With each attempt he appeared to retreat from his original boldness but, always, he generated new ideas and techniques that quantum theory later adopted. What appears to have finally convinced him of the correctness and deep significance of his quantum hypothesis (see Ref. 2) was its support of the definiteness of the statistical concept of entropy and the third law of thermodynamics.

It was during this period of doubt that Planck was editor of the *Annalen der Physik*. In 1905 he received Einstein's first relativity paper and stoutly defended Einstein's work. Thereafter he became one of young Einstein's patrons in scientific circles. However, he resisted for some time the very ideas on the quantum theory of radiation advanced by Einstein that subsequently confirmed and extended Planck's own work. Einstein, whose deep insight into electromagnetism and statistical mechanics was perhaps unequalled by anyone at the time, saw as a result of Planck's work the need for a sweeping change in classical statistics and electromagnetism. He advanced predictions and interpretations of many physical phenomena which were later strikingly confirmed by experiment. In the next chapter we turn to one of these phenomena and follow another road on the way to quantum theory.

QUESTIONS

1. Does a "black body" always appear black? Explain the term "black body."

2. The relation $R = \sigma T^4$ (Eq. 4-2) is exact for black bodies and holds for all temperatures. Why don't we use this relation then as the basis of a definition of temperature at, for instance, $100°C$?

3. (a) Compare the definitions and dimensions of spectral radiancy $\mathcal{R}$, radiancy R, and energy density ρ. (b) In Eq. 4-4, relating spectral radiancy and energy density, what dimensions would a proportionality constant need to have. The exact relation is $\mathcal{R}_T(\nu)\ d\nu = (c/4)\rho_T(\nu)\ d\nu$ (see Ref. 4).

4. Compare the radiation inside a cavity with the radiation emitted at the hole.

5. If we look into a cavity whose walls are kept at a constant temperature no details of the interior are visible. Explain.

6. "Pockets" formed by coals in a coal fire seem brighter than the coals themselves. Is the temperature in such pockets appreciably higher than the surface temperature of an exposed glowing coal?

7. A piece of metal glows with a bright red color at $1100°K$. At this same temperature, however, a piece of quartz does not glow at all. Explain. (*Hint.* Quartz is transparent to visible light.)

8. The law of equipartition of energy requires that the specific heat of gases be independent of the temperature (see *Physics*, Part 1, Sec. 23-8), in disagreement with experiment. Here we have seen that it leads to the Rayleigh-Jeans radiation law. How can you relate these two failures of the equipartition law?

9. Are there quantized quantities in classical physics? If so, give examples. Is energy quantized in classical physics?

10. Does it make sense to speak of charge quantization in physics? How, if at all, is this different from energy quantization?

11. Show that Planck's constant has the dimensions of angular momentum. Does this necessarily suggest that angular momentum is a quantized quantity?

12. For quantum effects to be "everyday" phenomena in our lives, what order of magnitude value would h need to have?

13. Would you expect atoms in a cavity wall to behave differently from atoms in a gas? Explain.

14. Does the quantization of the internal energy of an atom mean that the atom is a system obeying Planck's postulate? Explain.

15. Explain the relation between the rate at which electrons gain energy in the

Franck-Hertz arrangement and which upper energy state of a mercury atom can be excited. (*Hint.* Consider the mean free time between collisions.)

16. Discuss the remarkable fact that discreteness in energy was first found in analyzing a *continuous* spectrum (black body spectrum) rather than in a line (discrete) spectrum (mercury atom spectrum).

PROBLEMS

1. At what wavelength does a cavity radiator at 6000°K radiate most per unit wavelength?

2. (*a*) Assuming the surface temperature of the sun to be 5700°K, use the Stefan-Boltzmann law to determine the rest mass lost per second to radiation by the sun. Take the sun's diameter to be 1.4×10^9 m. (*b*) What fraction of the sun's rest mass is lost each year from electromagnetic radiation? Take the sun's rest mass to be 2.0×10^{30} kg.

3. Assuming that λ_{max} is in the near infrared for red heat and in the near ultraviolet for blue heat (see Fig. 4-7), approximately what temperature in Wien's displacement law corresponds to "red heat" . . . to "white heat" . . . to "blue heat"?

4. At what wavelength does the human body emit its maximum temperature radiation? List assumptions you make in arriving at an answer.

5. At a given temperature $\lambda_{max} = 6500$ Å for a black body cavity. What will λ_{max} be if the temperature of the cavity walls is increased so that the rate of emission of spectral radiation is doubled?

6. In a thermonuclear explosion the temperature in the fireball is momentarily 10^7°K. (*a*) Find the wavelength at which the radiation emitted is a maximum per unit wavelength. (*b*) What is the energy of a single quantity of radiation, called a photon, at this wavelength? (The energy, in Mev, of a photon of wavelength λ, in meters, is $(1.24 \times 10^{-12})/\lambda$.)

7. The *average* rate of solar radiation incident per unit area on the earth is 0.485 cal/cm²-min (or 355 watts/m²). (*a*) Explain the consistency of this number with the solar constant (the solar energy falling per unit time at normal incidence on unit area of the earth's surface) whose value is 1.94 cal/cm²-min (or 1340 watts/m²). (*b*) Consider the earth to be a black body radiating energy into space at this same rate. What surface temperature would the earth have under these circumstances?

8. Sirius, appearing to be the brightest star in the sky, has a temperature of about 11,000°C. What is its color?

9. Betelgeux and Rigel, the brightest stars in Orion, appear orange-red and white or slightly bluish, respectively. Using Fig. 4-7, compare their temperatures with that of the Sun.

10. Show that the Rayleigh-Jeans radiation law, Eq. 4-7, is not consistent with the Wien displacement law, $\nu_{max} \propto T$ (Eq. 4-3*a*) or $\lambda_{max}T = \text{const}$ (Eq. 4-3*b*).

★11. We obtain ν_{max} in the black body spectrum by setting $d\rho_T(\nu)/d\nu = 0$ and λ_{max} by setting $d\rho_T(\lambda)/d\lambda = 0$. Why can't we get from $\lambda_{max}T = \text{const}$ to $\nu_{max} = \text{const} \, xT$ simply by using $\lambda_{max} = c/\nu_{max}$? That is, why is it wrong to assume that $\nu_{max}\lambda_{max} = c$, where c is the speed of light?

12. A cavity radiator at 6000°K has a hole 0.10 mm in diameter drilled in its wall. (*a*) Find the power radiated through the hole in the wavelength range 5500–5510

Å. (*b*) Assuming that the radiation is delivered in small quantities of energy of $\Delta E = h\nu = hc/\lambda$, find the rate of emission of such so called photons.

★**13.** Use the relation $\mathfrak{R}_T(\nu)\,d\nu = \frac{c}{4}\rho_T(\nu)\,d\nu$ between spectral emissivity and energy density, together with Planck's radiation law, to derive Stefan's law. That is, show that

$$R_T = \int_0^\infty \frac{2\pi h}{c^2} \frac{\nu^3\,d\nu}{e^{h\nu/kT}-1} = \sigma T^4$$

where $\sigma = 2\pi^5 k^4/15c^2 h^3$.

$$\left(Hint. \quad \int_0^\infty \frac{q^3\,dq}{e^q-1} = \frac{\pi^4}{15}\right)$$

★**14.** Derive the Wien displacement law, $\lambda_{\max} T = 0.2014\ hc/k$, from Planck's radiation law by solving the equation

$$\frac{d}{d\lambda}\rho_T(\lambda) = 0.$$

(*Hint.* Set $hc/kT = x$ and show that the equation above leads to $e^{-x} + x/5 = 1$. Then show by substitution that $x = 4.965$ is the solution.)

15. (*a*) Show that at room temperature ($300°$K) the average translational kinetic energy of a molecule is about $\frac{1}{25}$ ev. (*b*) Assume that the first excited state of an atom is a few electron volts above its lowest (ground) state and show that a kinetic temperature of about 10^4°K is needed to excite an appreciable number of such atoms in a gas.

16. An atom, *at rest* and in its ground state, is struck by another atom of the same kind which is also in its ground state but has a translational kinetic energy K. Show, from conservation principles, that the collision must be elastic if $K < 2E$, where E is the first excitation energy of the atom (i.e., if K is less than twice the energy from the ground state to the first allowed excited state in the atom's energy level scheme).

17. (*a*) Let ΔE be the energy difference between the normal and first excited state of an atom. Show that the kinetic energy K of an electron, mass m, must be at least

$$K = \left(1 + \frac{m}{M}\right)\Delta E$$

to excite an atom of mass M. (*b*) Compare the relation of K to ΔE for a Franck-Hertz experiment (m ≪ M) to that for the situation in the previous problem ($m = M$).

18. The first excitation potential of sodium (potential through which electrons must fall to excite sodium) is 2.1 volts. Find the absolute temperature at which the average kinetic energy of sodium atoms equals the excitation energy.

19. In a Franck-Hertz kind of experiment, atomic hydrogen is bombarded with electrons and excitation potentials are found at 10.21 volts and 12.10 volts. (*a*) Explain the observation that three different lines of spectral emission accompany these excitations. (*Hint.* Draw an energy level diagram). (*b*) Now assume that the energy differences can be expressed as $h\nu$ and find the three allowed values of ν. (*c*) Assume that ν is the frequency of the emitted radiation and determine the wavelength of the observed spectral lines.

20. Assume, in the Franck-Hertz experiment, that the electromagnetic energy emitted by a Hg atom, in giving up the energy absorbed from 4.9 ev electrons, equals $h\nu$, where ν is the frequency corresponding to the 2536 Å mercury resonance line. Calculate the value of h according to the Franck-Hertz experiment and compare with Planck's value.

REFERENCES

1. Max Planck, "On the Theory of the Energy Distribution Law of the Normal Spectrum," *Annalen der Physik,* **4,** 533, 1901 [translated into English, in *Great Experiments in Physics,* edited by Morris H. Shamos, Hold Dryden, 1959].

2. Martin J. Klein, "Thermodynamics and Quanta in Planck's Work," *Physics Today,* **19,** 23, 1966.

3. Sir N. F. Mott, "On Teaching Quantum Phenomena," *Contemporary Physics,* August, 1964.

4. Richtmeyer, Kennard, and Lauritson, *Introduction to Modern Physics,* McGraw-Hill Book Co., 1955, Section 61.

CHAPTER 5

The Particle Nature of Radiation

5.1 Introduction

In this chapter we shall examine processes in which radiation interacts with matter. Three processes—the photoelectric effect, the Compton effect, and pair production—involve the absorption of radiation, and two processes—*bremsstrahlung* and pair annihilation—involve the production of radiation. In each case we obtain experimental evidence that radiation is particlelike in its interaction with matter as distinguished from the wavelike nature of radiation when it propagates.

5.2 The Photoelectric Effect

In 1886 and 1887, Heinrich Hertz performed the experiments that first confirmed the existence of electromagnetic waves and Maxwell's electromagnetic theory of light propagation. It is one of these fascinating and paradoxical facts in the history of science that in the course of his experiments Hertz noted the effect that Einstein later used to contradict other aspects of the classical electromagnetic theory. Hertz discovered that an electric discharge between two electrodes occurs more readily when ultraviolet light falls on one of the electrodes. It was shown soon after that the ultraviolet light facilitates the discharge by causing electrons to be emitted from the cathode surface. The ejection of electrons from a surface by the action of light is called the *photoelectric effect*.

Figure 5-1 shows an apparatus used to study the photoelectric effect. A glass envelope encloses the apparatus in an evacuated space. Monochromatic light, incident through a quartz window, falls on the metal plate A and liberates electrons, called *photoelectrons*. The electrons can be detected as a current if they are attracted to the metal cup B by means of a potential difference V applied between A and B. The galvanometer G serves to measure this photoelectric current.

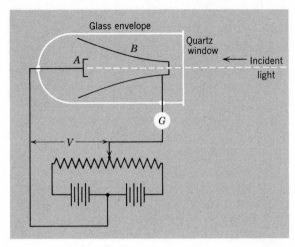

Figure 5-1. An apparatus used to study the photoelectric effect. *V* can be varied continuously through both positive and negative values.

Figure 5-2 (curve *a*) is a plot of the photoelectric current, in an apparatus like that of Fig. 5-1, as a function of the potential difference *V*. If *V* is made large enough, the photoelectric current reaches a certain limiting (saturation) value at which all photoelectrons ejected from *A* are collected by cup *B*.

If *V* is reduced to zero and gradually reversed in sign, the photoelectric

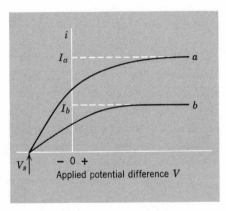

Figure 5-2. Graphs of current *i* as a function of voltage *V* from data taken with the apparatus of Fig. 5-1. The applied potential difference *V* is called positive when the cup *B* in Fig. 5-1 is positive with respect to the photoelectric surface *A*. In curve *b* the incident light intensity has been reduced to one half that of curve *a*. The stopping potential V_s is independent of light intensity but the saturation currents I_a and I_b are directly proportional to it.

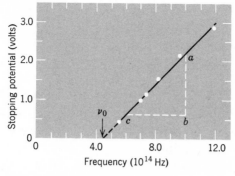

Figure 5-3. A plot of Millikan's measurements of the stopping potential at various frequencies for sodium. The cutoff frequency ν_0 is 4.39×10^{14} Hz.

current does not drop to zero, which suggests that the electrons are emitted from A with kinetic energy. Some will reach cup B in spite of the fact that the electric field opposes their motion. However, if this reversed potential difference is made large enough, a value V_s (the *stopping potential*) is reached at which the photoelectric current does drop to zero. This potential difference V_s, multiplied by electron charge, measures the kinetic energy K_{max} of the *fastest* ejected photoelectron. That is,

$$K_{max} = eV_s. \tag{5-1}$$

K_{max} turns out experimentally to be independent of the intensity of the light, as is shown by curve b in Fig. 5-2 in which the light intensity has been reduced to one-half the value used in obtaining curve a.

Suppose now that we vary the frequency, or wavelength, of the incident light. Figure 5-3 shows the stopping potential V_s as a function of the frequency of the light incident on a clean sodium surface. Note that there is a definite *cutoff* frequency ν_0, below which no photoelectric effect occurs. These data were taken in 1914 by R. A. Millikan whose painstaking work on the photoelectric effect won him the Nobel prize in 1923. Because the photoelectric effect for visible or near-visible light is largely a surface phenomenon, it is necessary to avoid oxide films, grease, or other surface contaminants. Millikan devised a technique to cut shavings from the metal surface under vacuum conditions, a "machine shop in vacuo" as he called it.

There are three major features of the photoelectric effect that cannot be explained in terms of the classical wave theory of light.

1. Wave theory requires that the oscillating electric vector $\mathbf{E}$ of the light wave increase in amplitude as the intensity of the light beam is increased. Since the force applied to the electron is $e\mathbf{E}$, this suggests that *the kinetic energy of the photoelectrons should also increase as the light beam is made more intense*. However, Fig. 5-2 shows that $K_{max}(= eV_s)$ *is independent of the light intensity;* this has been tested over a range of intensities of 10^7.

2. According to the wave theory the photoelectric effect should occur for

any frequency of the light, provided only that the light is intense enough to provide the energy needed to eject the photoelectrons. However, Fig. 5-3 shows that there exists, for each surface, *a characteristic cutoff frequency* ν_0. *For frequencies less than* ν_0, *the photoelectric effect does not occur, no matter how intense the illumination.*

3. If the energy acquired by a photoelectron is absorbed directly from the wave incident on the metal plate, the "effective target area" for an electron in the metal is limited and probably not much more than that of a circle of the order of an atomic diameter. In the classical theory the light energy is uniformly distributed over the wave front. Thus, if the light is feeble enough, there should be a measurable time lag, which we shall estimate in Example 1, between the impinging of the light on the surface and the ejection of the photoelectron. During this interval the electron should be absorbing energy from the beam until it had accumulated enough to escape. *However, no detectable time lag has ever been measured.* This disagreement is particularly striking when the photoelectric substance is a gas; under these circumstances collective absorption mechanisms can be ruled out and the energy of the emitted photoelectron must certainly be soaked out of the light beam by a single atom or molecule.

◆ **Example 1.** A foil of potassium is placed 3 meters from a weak light source whose power is 1.0 watt. Assume that an ejected photoelectron may collect its energy from a circular area of the foil whose radius is, say, one atomic radius ($r \simeq 0.5 \times 10^{-10}$ meter). The energy required to remove an electron through the potassium surface is about 1.8 ev; how long would it take for such a "target" to absorb this much energy from such a light source? Assume the light energy to be spread uniformly over the wave front.

The target area is $\pi(0.5 \times 10^{-10}$ meter$)^2$; the area of a 3-meter sphere centered on the light source is $4\pi(3$ meters$)^2$. Thus if the light source radiates uniformly in all directions—that is, if the light energy is uniformly distributed over spherical wavefronts spreading out from the source, in agreement with classical theory—the rate P at which energy falls on the target is given by

$$P = (1.0 \text{ watt})\left(\frac{(\pi/4) \times 10^{-20} \text{ meter}^2}{36\pi \text{ meter}^2}\right) = 7 \times 10^{-23} \text{ joule/sec.}$$

Assuming that all this power is absorbed, we may calculate the time required for the electron to acquire enough energy to escape; we find

$$t = \left(\frac{1.8 \text{ ev}}{7 \times 10^{-23} \text{ joule/sec}}\right)\left(\frac{1.6 \times 10^{-19} \text{ joule}}{1 \text{ ev}}\right) \simeq 4000 \text{ secs}$$

Of course, we could modify the above picture to reduce the calculated time by assuming a much larger effective target area. The most favorable assumption, that energy is transferred by a resonance process from light wave to electron, leads to a target area of λ^2, where λ is the wavelength of the light. But we would still obtain a finite time lag that is within our ability to measure experimentally. (For ultraviolet light of $\lambda = 100$ Å, for example, $t \simeq 1$ second). However, no time lag has been detected under any circumstances, the early experiments setting an upper limit of 10^{-9} sec on any such possible delay! ◆

5.3 Einstein's Quantum Theory of the Photoelectric Effect

In 1905, many years before Millikan's experiments were performed, Einstein called into question the classical theory of light, proposed a new theory, and cited the photoelectric effect as one application that could test which theory was correct (see Ref. 1). Planck had, at first, restricted his concept of energy quantization to the emission or absorption mechanism of a material oscillator. He believed that light energy, once emitted, was distributed in space like a wave. Einstein proposed instead that the radiant energy itself existed in concentrated bundles which later came to be called *photons*. The energy E of a single photon is given by

$$E = h\nu \tag{5-2}$$

where ν is the frequency of the radiation and h is Planck's constant. In effect, Einstein proposed a granular structure to radiation itself, radiant energy being distributed in space in a discontinuous way. This suggests that light, though wavelike in its propagation, could behave like corpuscles in emission and absorption processes, i.e., that it can be emitted or absorbed in individual bundles, or photons. Millikan, whose brilliant experiments over many years later verified Einstein's ideas in every detail "contrary to my own expectations," spoke of Einstein's "bold, not to say reckless hypothesis."

Applying the photon concept to the photoelectric effect, Einstein proposed that the entire energy of a photon is transferred to a single electron in a metal. When the electron is emitted from the surface of the metal, then, its kinetic energy will be

$$K = h\nu - w \tag{5-3}$$

where $h\nu$ is the energy of the absorbed incident photon and w is the work required to remove the electron from the metal. This work is needed to overcome the attractive fields of the atoms in the surface and losses of kinetic energy caused by internal collisions of the electron. Some electrons are bound more tightly than others; some lose energy in collisions on the way out. In the case of loosest binding and no internal losses the photoelectron will emerge with the maximum kinetic energy, K_{max}. Hence,

$$K_{max} = h\nu - w_0 \tag{5-4}$$

where w_0, a characteristic energy of the metal called the *work function*, is the minimum energy needed by an electron to pass through the metal surface.

Consider now how Einstein's photon hypothesis meets the three objections raised against the wave theory interpretation of the photoelectric effect. As for objection 1 (the lack of dependence of K_{max} on the intensity of illumination), there is complete agreement of the photon theory with experiment. Doubling the light intensity merely doubles the number of photons and thus doubles the photoelectric current; it does *not* change the energy distribution of the individual electrons, the energy ($= h\nu$) of the individual photons, or the nature of the individual photoelectric process described by Eq. 5-3.

Objection 2 (the existence of a cutoff frequency) follows at once from Eq. 5-4. If $K_{\max}$ equals zero we have

$$h\nu_0 = w_0 \tag{5-5}$$

which asserts that a photon of frequency ν_0 has just enough energy to eject the photoelectrons and none extra to appear as kinetic energy. If the frequency is reduced below ν_0, the individual photons, regardless of how many of them there are (that is, no matter how intense the illumination), will not have enough energy individually to eject photoelectrons.

Objection 3 (the absence of a time lag) follows from the photon theory because the required energy is supplied in concentrated bundles. It is *not* spread uniformly over a large area, as we assumed in Example 1, which is based on the assumption that the classical wave theory is true. If there is any illumination at all incident on the cathode, then there will be at least one photon that hits it; this photon will be immediately absorbed, by *some* atom, leading to the immediate emission of a photoelectron.

Let us rewrite Einstein's photoelectric equation (Eq. 5-4) by substituting eV_s for $K_{\max}$ (see Eq. 5-1). This yields

$$V_s = \left(\frac{h}{e}\right)\nu - (w_0/e).$$

Thus Einstein's theory predicts a linear relationship between the stopping potential V_s and the frequency ν, in complete agreement with experiment (see Fig. 5-3). The slope of the experimental curve in this figure should be h/e, or

$$\frac{h}{e} = \frac{ab}{bc} = \frac{2.20 \text{ volts} - 0.65 \text{ volt}}{10.0 \times 10^{14}/\text{sec} - 6.0 \times 10^{14}/\text{sec}} = 3.9 \times 10^{-15} \text{ volt-sec.}$$

We can find h by multiplying this ratio by the electronic charge e; $h = (3.9 \times 10^{-15} \text{ volt-sec})(1.6 \times 10^{-19} \text{ coul}) = 6.2 \times 10^{-34} \text{ joule-sec.}$ From a more careful analysis of these and other data, including data taken with lithium surfaces, Millikan found the value $h = 6.57 \times 10^{-34}$ joule-sec, with an accuracy of about 0.5%. This early measurement was in good agreement with the value of h derived from Planck's radiation formula. This numerical agreement in two determinations of h, using completely different phenomena and theories, is striking. The modern value of h, deduced from diverse experiments, is

$$h = 6.62517 \times 10^{-34} \text{ joule-sec.}$$

To quote Millikan: "The photoelectric effect . . . furnishes a proof which is quite independent of the facts of black-body radiation of the correctness of the fundamental assumption of the quantum theory, namely, the assumption of a discontinuous of explosive emission of the energy absorbed by the electronic constituents of atoms from . . . waves. It materializes, so to speak, the quantity h discovered by Planck through the study of black body radiation and gives us a confidence inspired by no other type of phenomenon that the primary physical conception underlying Planck's work corresponds to reality."

▶ **Example 2.** Deduce the work function for sodium from Fig. 5-3.

The intersection of the straight line in Fig. 5-3 with the horizontal axis is the cutoff frequency, $\nu_0 = 4.39 \times 10^{14}/\text{sec}$. Substituting this into Eq. 5-5 gives us

$$w_0 = h\nu_0 = (6.63 \times 10^{-34} \text{ joule-sec})(4.39 \times 10^{14}/\text{sec})$$

$$= 2.92 \times 10^{-19} \text{ joule} \left(\frac{1 \text{ ev}}{1.60 \times 10^{-19} \text{ joule}} \right)$$

$$= 1.82 \text{ ev}.$$

Note that the work function w_0 also can be obtained directly from Fig. 5-3 as the magnitude of the negative intercept of the extended straight line with the vertical axis.

For most conducting metals the value of the work function is of the order of a few electron volts. It is the same as the work function for thermionic emission from these metals.

Example 3. At what rate per unit area do photons strike the metal plate in Example 1? Assume that the light is monochromatic, of wavelength 5890 Å (sodium light).

The rate per unit area which energy falls on a metal plate 3 meters from the light source (see Example 1) is

$$R = \frac{1.0 \text{ watt}}{100 \text{ m}^2} = 10^{-2} \text{ joule/m}^2 \text{ sec} \left(\frac{1 \text{ ev}}{1.6 \times 10^{-19} \text{ joule}} \right)$$

$$= 6.3 \times 10^{+16} \text{ ev/m}^2 \text{ sec}$$

Each photon has an energy of

$$E = h\nu = \frac{hc}{\lambda} = \frac{(6.63 \times 10^{-34} \text{ joule/sec})(3.00 \times 10^8 \text{ m/sec})}{(5.89 \times 10^{-7} \text{ m})}$$

$$= 3.4 \times 10^{-19} \text{ joule} \left(\frac{1 \text{ ev}}{1.6 \times 10^{-19} \text{ joule}} \right)$$

$$= 2.1 \text{ ev}.$$

Thus the rate R at which photons strike the plate is

$$R = (6.3 \times 10^{16} \text{ ev/m}^2 \text{ sec}) \left(\frac{1 \text{ photon}}{2.1 \text{ ev}} \right) = 3 \times 10^{16} \frac{\text{photons}}{\text{m}^2 \text{ sec}}.$$

The photoelectric effect will occur in this case because the photon energy (2.1 ev) is greater than the work function for the surface (1.82 ev; see example 2). Note that if the wavelength is sufficiently increased (i.e., if ν is sufficiently decreased) the photoelectric effect will not occur, no matter how large the rate R might be.

This example suggests that the intensity of light I can be regarded as the product of N, the number of photons per unit area per unit time, and $h\nu$, the energy of a single photon. We see that even at the relatively low intensity here (10^{-2} watt/m^2) the number N is extremely large ($\simeq 10^{16}$ photons/m^2 — sec) so that the energy of any one photon is very small. This accounts for the extreme fineness of the granularity of radiation and suggests why ordinarily it is difficult to detect at all. It is analogous to detecting the atomic structure of bulk matter which for most purposes can be regarded as continuous, the discreteness being revealed only under special circumstances. ◀

It was in 1921 that Einstein received the Nobel prize for his discovery of the law of the photoelectric effect. Before Millikan's complete experimental validation of this law, Einstein was recommended to membership in the Prussian Academy of Science by Planck and others. Their early negative attitude toward the light quantum hypothesis is revealed in their signed affidavit, praising Einstein, in which they wrote: "Summing up, we may say that there is hardly one among the great problems, in which modern physics is so rich, to which Einstein has not made an important contribution. That he may have sometimes missed the target in his speculations, as, for example, in his hypothesis of light quanta, cannot really be held too much against him, for it is not possible to introduce fundamentally new ideas, even in the most exact sciences, without occasionally taking a risk."

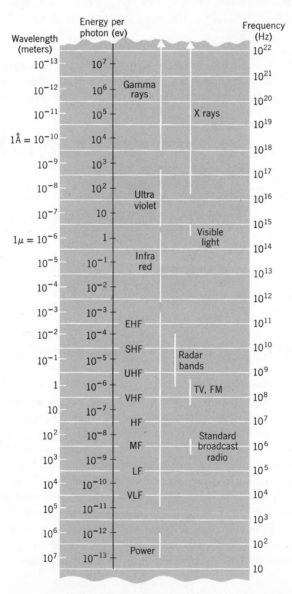

Figure 5-4. The electromagnetic spectrum, showing wavelength, frequency, and energy per photon on a logarithmic scale.

Today the photon hypothesis is used throughout the electromagnetic spectrum, not only in the light region (see Fig. 5-4). A microwave cavity, for example, can be said to contain photons. At $\lambda = 10$ cm, a typical microwave wavelength, the photon energy can be computed as above to be 1.20×10^{-5} ev. This quantum energy is much too low to eject photoelectrons from metal surfaces. For X rays, or for energetic gamma rays such as are emitted from radioactive nuclei, the photon energy may be 1 Mev or higher. Such photons can eject electrons bound deep in heavy atoms by energies of the order of 10^5 ev. The photons in the visible region of the electromagnetic spectrum are not energetic enough to do this, the photoelectrons which they eject being the so-called *conduction* electrons which are bound to the metal by energies of only a few electron volts.

Notice that the photons are absorbed in the photoelectric process. This requires the electrons to be bound to atoms, or solids, for a truly free electron cannot absorb a photon and conserve both energy and momentum in the process (see Prob. 17). We must have a bound electron, therefore, in which case the binding forces serve to transmit momentum to the atom or solid. Because of the large mass of atom, or solid, compared to the electron, the system can absorb a large amount of momentum without, however, acquiring a significant amount of energy. Our photoelectric energy equation remains valid, the effect being possible only because there is a heavy recoiling particle in addition to an ejected electron. The photoelectric effect is one important way in which photons, of energy up to and including X-ray energies, are absorbed by matter. At higher energies other photon absorption processes, soon to be discussed, become more important.

Finally, it should be emphasized here that in the Einstein picture a quantum of frequency ν has exactly the energy $h\nu$; it does *not* have energies that are integral multiples of $h\nu$. Of course, there can be n photons of frequency ν so that the energy at that frequency can be $nh\nu$. In treating cavity radiation in the Einstein picture, however, one deals with a "photon gas," for the radiant energy is spatially localized in bundles rather than spatially extending in standing waves. Years after the Planck deduction of the cavity radiation formula, Bose and Einstein derived the same formula on the basis of a photon gas. They used a form of quantum statistics and, in so doing, eliminated the last classical vestige underlying the Planck derivation, namely the classical Boltzmann distribution.

5.4 The Compton Effect

The corpuscular (particlelike) nature of radiation received dramatic confirmation in 1923 from the experiments (Ref. 2) of A. H. Compton. Compton caused a beam of X rays of sharply defined wavelength λ to fall on a graphite target, as shown in Fig. 5-5. For various angles of scattering, he measured the intensity of the scattered X rays as a function of their wavelength. Figure 5-6 shows his experimental results. We see that, although the incident beam consists essentially of a single wavelength λ, the scattered X rays have intensity peaks at *two* wavelengths; one of them is the same as the incident wavelength,

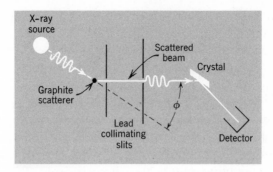

Figure 5-5. Compton's experimental arrangement. Monochromatic X rays of wavelength λ fall on a graphite scatterer. The distribution of intensity with wavelength is measured for X rays scattered at any selected angle ϕ. The scattered wavelengths are measured by observing Bragg reflections from a crystal. Their intensities are measured by a detector, such as an ionization chamber.

the other, λ', being larger by an amount $\Delta\lambda$. This so-called *Compton shift*, $\Delta\lambda = \lambda' - \lambda$, varies with the angle at which the scattered X rays are observed.

The presence of scattered wavelength λ' cannot be understood if the incident x-radiation is regarded as a classical electromagnetic wave. In the classical model the oscillating electric field vector in the incident wave of frequency ν acts on the free electrons in the scattering block and sets them oscillating at that same frequency. These oscillating electrons, like charges surging back and forth in a small radio transmitting antenna, radiate electromagnetic waves that again have this same frequency ν. Hence, in the wave picture the scattered wave should have the same frequency ν and the same wavelength λ as the incident wave.

Compton (and independently P. Debye) interpreted his experimental results by postulating that the incoming X ray beam was not a wave of frequency ν but a collection of photons, each of energy $E = h\nu$, and that these photons collided with free electrons in the scattering block like a collision between billiard balls. In this view, the "recoil" photons emerging from the target make up the scattered radiation. Since the incident photon transfers some of its energy to the electron with which it collides, the scattered photon must have a lower energy E'; it must therefore have a lower frequency $\nu'(= E'/h)$, which implies a larger wavelength $\lambda'(= c/\nu')$. This point of view accounts qualitatively for the wavelength shift, $\Delta\lambda = \lambda' - \lambda$. Notice that in the interaction the X rays are regarded as particles, not as waves, and that, as distinguished from their behavior in the photoelectric process, the X-ray photons are scattered rather than absorbed. Let us now analyze a single photon-electron collision quantitatively.

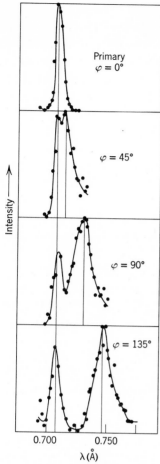

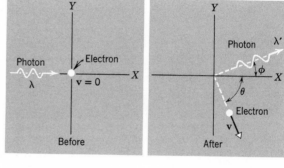

Figure 5-6. Compton's experimental results. The solid vertical line on the left corresponds to the wavelength λ, that on the right to λ'. Results are shown for four different angles of scattering ϕ. Note that the Compton shift, $\Delta\lambda = \lambda' - \lambda$, for $\phi = 90°$ is $h/m_0 c = 0.0243$ Å.

Figure 5-7. Compton's interpretation. A photon wavelength λ is incident on a free electron at rest. On collision, the photon is scattered at an angle ϕ with increased wavelength λ', while the electron moves off with speed $\mathbf{v}$ at angle θ.

Figure 5-7 shows a collision between a photon and an electron, the electron assumed to be initially at rest and essentially free, that is, not bound to the atoms of the scatterer. Let us apply the laws of conservation of mass-energy and of linear momentum to this collision. We use relativistic expressions because the recoil speed v of the electrons may be near the speed of light.

Conservation of mass-energy yields

$$h\nu = h\nu' + (m - m_0)c^2$$

in which $h\nu$ is the energy of the incident photon, $h\nu'$ is the energy of the scattered photon, and $(m - m_0)c^2$ is the kinetic energy acquired by the recoiling electron, initially at rest. With $m = m_0/\sqrt{1 - \beta^2}$, $\nu = c/\lambda$ and $\nu' = c/\lambda'$, this expression becomes

$$\frac{hc}{\lambda} = \frac{hc}{\lambda'} + m_0 c^2 \left(\frac{1}{\sqrt{1 - v^2/c^2}} - 1 \right). \tag{5-6}$$

Now let us apply the law of conservation of linear momentum. The momentum of a photon is given by $p = E/c$, an expression derivable from relativity theory (see Sec. 3.5) or from electromagnetic theory (see *Physics, Part II*, Sec. 40.2). Using $E = h\nu$ we obtain $p = h\nu/c = h/\lambda$. There are two components of momentum in the plane of the scattered photon and electron (see Fig. 5-7); the conservation of the x-component of linear momentum gives us

$$\frac{h}{\lambda} = \frac{h}{\lambda'} \cos \phi + \frac{m_0 v}{\sqrt{1 - v^2/c^2}} \cos \theta, \tag{5-7}$$

and the conservation of the y-component gives us

$$0 = \frac{h}{\lambda'} \sin \phi - \frac{m_0 v}{\sqrt{1 - v^2/c^2}} \sin \theta. \tag{5-8}$$

Our immediate goal is to find $\Delta\lambda = \lambda' - \lambda$, the wavelength shift of the scattered photons, so that we may compare that expression with the experimental results of Fig. 5-6. Compton's experiment did not involve observations of the recoil electron in the scattering block. Of the five variables (λ, λ', v, ϕ, and θ) that appear in the three equations (5-6, 5-7, 5-8), we may eliminate two by combining the equations. We choose to eliminate v and θ, which deal only with the electron, thereby reducing the three equations to a single relation among the variables λ, λ', and ϕ. The result (see Prob. 18) is

$$\Delta\lambda(= \lambda' - \lambda) = \frac{h}{m_0 c}(1 - \cos \phi). \tag{5-9}$$

Notice that $\Delta\lambda$, *the Compton shift*, depends only on the scattering angle ϕ and *not* on the initial wavelength λ. The *change* in wavelength is independent of the incident wavelength. The constant $h/m_0 c$, called the Compton wavelength, has the value 0.0243 Å. Equation 5-9 predicts the experimentally observed Compton shifts of Fig. 5-6 within the experimental limits of error. The shift was found to be independent of the material of the scatterer and of the incident wavelength, as required by Compton's interpretation, and to be proportional to $(1 - \cos \phi)$. Note that $\Delta\lambda$, in the equation, varies from zero (for $\phi = 0$, corresponding to a "grazing" collision in Fig. 5-7, the incident photon being scarcely deflected) to $2h/m_0c = 0.049$ Å (for $\phi = 180°$, corresponding to a "head-on" collision, the incident photon being reversed in direction). In Fig. 5-8 we plot $\Delta\lambda$ versus ϕ. Subsequent experiments (by Compton, Simon, Wilson, Bothe, Geiger, Blass) detected the recoil electron in the process, showed that it appeared simultaneously with the scattered X ray, and confirmed quantitatively the predicted electron energy and direction of scattering. (See Probs. 19 and 20.)

It remains to explain the presence of the peak in Fig. 5-6 for which the photon wavelength does *not* change on scattering. We refer to this as the *unmodified* line, the Compton shifted line being called the *modified* one. In our equations, we assumed that the electron with which the photon collides is free. Even when the electron is initially bound this assumption can be justified as approximately correct if the kinetic energy acquired by the electron is much larger than its binding energy. However, if the electron is strongly bound to an atom or if the incident photon energy is small, there is a chance that the electron will not be ejected from the atom, in which case, the collision can be regarded as taking place between the photon and the whole atom. The ionic core, to which the electron is bound in the scattering target, recoils as a whole during such a collision. In that case the mass M_0 of the atom is the effective mass and it must be substituted in Eq. 5-9 for the electron mass m_0. Since $M_0 \gg m_0$ ($M_0 \simeq 22{,}000\ m_0$ for carbon, for instance), the Compton shift for collisions with tightly bound electrons is seen to be immeasurably small (one millionth of an angstrom unit for the carbon atom), so that the scattered photon is observed to be unmodified in wavelength. Some photons then are scattered from electrons which are freed by the collision; these photons are modified in wavelength. Other photons are scattered from electrons which remain bound during the collision; these photons are not modified in wavelength.

The breadth of the peaks in Fig. 5-6 also can be simply accounted for. It is caused by the motion of the electrons, which were assumed initially to

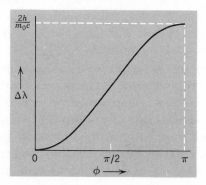

Figure 5-8. A plot of the Compton shift $\Delta\lambda$ versus scattering angle ϕ, illustrating Compton's theoretical results, $\Delta\lambda = (h/m_0c)(1 - \cos\phi)$.

be at rest in our analysis. If we include in our calculation the components of the velocities of the atomic electrons in (or opposite to) the direction of the incident radiation, the observed line broadening can be deduced.

We saw earlier in the cavity radiation discussion, and also in the photoelectric discussion, that Planck's constant h is a measure of the granularity or discreteness in energy. Classical physics corresponds to $h = 0$, since in this case all energy spectra would be continuous. Notice that here in the Compton effect, Planck's constant h again plays a central role. If h were equal to zero there would be no Compton effect, for then $\Delta\lambda = 0$ and classical theory would be valid. The quantity h is the central constant of quantum physics.

The fact that h is not zero means that classical physics is not valid in general; but the fact that h is very small often makes quantum effects difficult to detect. For example, the quantity h/m_0c in the Compton formula has the value 0.0243 Å when the scatterer is a free electron. But if m_0 is the mass of an atom, not to mention bulk matter, h/m_0c is already at least 2000 times smaller and virtually undetectable. Hence, as $m_0 \to \infty$ the quantum scattering result merges with the classical result, the scattered radiation then having the same frequency as the incident radiation. It is in the atomic and subatomic domain, where m_0 is small, that the classical results fail.

Similarly, if the incident radiation were in the visible, microwave, or radio part of the electromagnetic spectrum, then λ would be very large compared to $\Delta\lambda$, the Compton shift. The scattered radiation would be measured to have the same frequency as the incident radiation within the experimental limits. For microwaves of, say, $\lambda = 10$ cm we have $\Delta\lambda/\lambda = 0.0243$ Å/10 cm = 2×10^{-11} whereas for X rays of say, $\lambda = 1$ Å we have $\Delta\lambda/\lambda = 0.0243$ Å/ 1 Å = 2×10^{-2}. It is no accident that the Compton effect was discovered in the X ray region, for the quantum nature of radiation reveals itself to us at short wavelengths. As $\lambda \to \infty$ the quantum results merge with the classical. It is in the short wavelength region, where λ is small, that the classical results fail. Recall the "ultraviolet catastrophe" of classical physics, wherein classical predictions of black body radiation agreed with experiment at long wavelengths but diverged radically at short wavelengths. From the point of view of the energy quantum $h\nu$, this is best intrepreted as due to the smallness of h. For at long wavelengths, the frequency ν is small and the granularity in energy, $h\nu$, is so small as to be virtually indistinguishable from a continuum. But at short wavelengths, where ν is large, $h\nu$ is no longer too small to be detected and quantum effects abound. *

In his paper (Ref. 4) on "A Quantum Theory of the Scattering of X-rays by Light Elements," Compton wrote: "The present theory depends essentially upon the assumption that each electron which is effective in the scattering scatters a complete quantum. It involves also the hypothesis that the quanta of radiation are received from definite directions and are scattered in definite directions. The experimental support of the theory indicates very convincingly that a radiation quantum carries with it directed momentum as well as energy."

---

*See Ref. 3 for a delightful popularization of a world in which the physical constants c, G, and h make themselves obvious.

The need for a quantum, or particle, interpretation of processes dealing with the interaction between radiation and matter seemed clear, but at the same time a wave theory of radiation seemed necessary to understand interference and diffraction phenomena. Clearly the idea that radiation is neither purely a wave phenomenon nor merely a stream of particles has to be considered seriously. But whatever radiation is, we have seen that it behaves wavelike under some circumstances and particlelike under other circumstances. Indeed, the paradoxical situation is revealed most forcefully in Compton's very experiment where (a) a crystal spectrometer is used to measure X-ray wavelengths, the measurement being interpreted by a wave theory of diffraction and (b) the scattering affects the wavelength in a way that can be understood only by treating the X rays as particles. It is in the very expressions $E = h\nu$ and $p = h/\lambda$ that the waves attributes (ν and λ) and the particle attributes (E and p) are combined.

▶ **Example 4.** Consider (I) an X-ray beam, with $\lambda = 1.00$ Å and (II) a gamma-ray beam from a Cs^{137} sample, with $\lambda = 1.88 \times 10^{-2}$ Å. If the radiation scattered from free electrons is viewed at 90° to the incident beam: (a) what is the Compton wavelength shift in each case?; (b) what kinetic energy is given to a recoiling electron in each case?; (c) what percentage of the incident photon energy is lost in the collision in each case?

(a) The Compton shift, with $\phi = 90°$, is

$$\Delta\lambda = \frac{h}{m_0 c}(1 - \cos\phi) = \frac{6.63 \times 10^{-3} \text{ joule/sec}}{(9.11 \times 10^{-31} \text{ kg})(3.00 \times 10^8 \text{ m/sec})}(1 - \cos 90°)$$

$$= 2.43 \times 10^{-12} \text{ m} = 0.0243 \text{ Å}.$$

This result is independent of the incident wavelength, the same for the gamma rays as the X rays.

(b) Let K be the kinetic energy of the recoiling electron, so that Eq. 5-6 can be written as

$$\frac{hc}{\lambda} = \frac{hc}{\lambda'} + k.$$

Then, since $\lambda' = \lambda + \Delta\lambda$, we have

$$\frac{hc}{\lambda} = \frac{hc}{\lambda + \Delta\lambda} + K$$

so that

$$K = \frac{hc\,\Delta\lambda}{\lambda(\lambda + \Delta\lambda)}.$$

For the X-ray beam, with $\lambda = 1.00$ Å, we have

$$K = \frac{(6.63 \times 10^{-34} \text{ joule/sec})(3.00 \times 10^8 \text{ m/sec})(2.43 \times 10^{-12} \text{ m})}{(1.00 \times 10^{-10} \text{ m})(1.00 + 0.024) \times 10^{-10} \text{ m}}$$

$$= 4.73 \times 10^{-17} \text{ joule} = 295 \text{ ev}.$$

For the gamma-ray beam, with $\lambda = 1.88 \times 10^{-2}$ Å, we have

$$K = \frac{(6.63 \times 10^{-34} \text{ joule/sec})(3.00 \times 10^8 \text{ m/sec})(2.43 \times 10^{-12} \text{ m})}{(1.88 \times 10^{-12} \text{ m})(0.0188 + 0.0243) \times 10^{-10} \text{ m}}$$

$$= 5.98 \times 10^{-14} \text{ joule} = 378 \text{ kev}.$$

(*c*) The incident X-ray photon energy is

$$E = h\nu = \frac{hc}{\lambda} = \frac{(6.63 \times 10^{-34} \text{ joule/sec})(3.00 \times 10^8 \text{ m/sec})}{1.00 \times 10^{-10} \text{ m}}$$

$$= 1.99 \times 10^{-15} \text{ joule} = 12,400 \text{ ev} = 12.4 \text{ kev}.$$

The energy lost by the photon equals that gained by the electron, or 295 ev, so that the percentage loss in energy is

$$\frac{295 \text{ ev}}{12,400 \text{ ev}} \times 100\% = 2.4\%$$

The incident gamma-ray photon energy is

$$E = h\nu = \frac{hc}{\lambda} = \frac{(6.63 \times 10^{-34} \text{ joule/sec})(3.00 \times 10^8 \text{ m/sec})}{1.88 \times 10^{-12} \text{ m}}$$

$$= 1.06 \times 10^{-13} \text{ joule} = 660 \text{ kev}.$$

The energy lost by the photon equals that gained by the electron, or 378 kev, so that the percentage loss in energy is

$$\frac{378 \text{ kev}}{660 \text{ kev}} \times 100\% = 57\%$$

Hence, the more energetic photons (which have small wavelengths) experience a larger *percent* loss in energy in Compton scattering. This corresponds to the fact that the photons of smaller wavelengths experience a larger *percent* increase in wavelength on being scattered. This becomes clear from the expression for fractional loss in energy, given simply by

$$\frac{K}{E} = \frac{hc\,\Delta\lambda/\lambda(\lambda + \Delta\lambda)}{hc/\lambda} = \frac{\Delta\lambda}{\lambda + \Delta\lambda}.$$

From this it can be shown that at $\lambda = 2500$ Å, for example, the percentage loss (for $\phi = 90°$) is less than one hundredth of one percent, whereas at $\lambda = 1.25 \times 10^{-2}$ Å, corresponding to gamma rays of 1 Mev, the percentage loss (for $\phi = 90°$) is 67%. ◄

5.5 Photons and X-ray Production

X rays, so named by their discoverer Roentgen because their nature was then unknown, are radiations in the electromagnetic spectrum of wavelength less than about 1.0 Å. They show the typical transverse wave behavior of polarization, interference and diffraction that is found in light and all other electromagnetic radiation. X rays are produced in the target of an X-ray tube, illustrated in Fig. 5-9a, when a beam of energetic electrons, accelerated through a potential difference of thousands of volts, is stopped upon striking the target. According to classical physics, the deceleration of the electrons, brought to rest in the target material, results in the emission of a continuous spectrum of electromagnetic radiation.

Figure 5-9b shows, for four different values of the incident electron energy, how the X rays emerging from a tungsten target are distributed in wavelength. (In addition to the continuous X-ray spectrum, X-ray lines characteristic of the target material are emitted. We discuss these later.) The most notable feature of these curves is that, for a given electron energy, there exists a well-defined minimum wavelength λ_{min}; for 40 kev electrons, for instance,

λ_{min} is 0.311 Å. Although the overall shape of the continuous X-ray distribution spectrum depends on the choice of target material as well as on the electron accelerating potential V, the value of λ_{min} depends only on V, being the same for all target materials. Classical electromagnetic theory cannot account for this fact, there being no reason why waves whose length is less than a certain critical value should not emerge from the target.

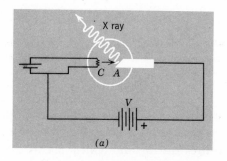

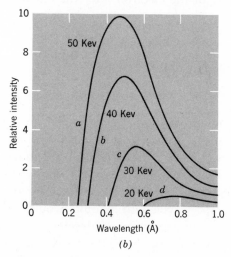

Figure 5-9. (*a*) Electrons are emitted thermionically from the heated cathode filament C and are accelerated toward the anode target A by the applied potential V. X rays are emitted from the target when electrons are stopped in striking it. (*b*) Plots of intensity versus wavelength of the continuous X-ray spectrum emitted from a tungsten target for four different values of ev, the incident electron energy.

A ready explanation appears, however, if we regard the X-rays as photons. Figure 5-10 shows the elementary process that, on the photon view, is responsible for the continuous X-ray spectrum of Fig. 5-9*b*. An electron of initial kinetic energy K is decelerated during an encounter with a heavy target nucleus, the energy it loses appearing in the form of radiation as an X-ray photon. The electron interacts with the charged nucleus via the Coulomb

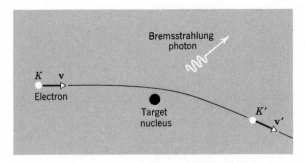

Figure 5-10. The bremsstrahlung process responsible for the production of X rays in the continuous spectrum.

field, transferring momentum to the nucleus. The accompanying deceleration of the electron leads to photon emission. The target nucleus is so massive that the energy it acquires during the collision can safely be neglected. If K' is the kinetic energy of the electron after the encounter, then the energy of the photon is

$$h\nu = K - K'$$

and the photon wavelength follows from

$$hc/\lambda = K - K' \tag{5-10}$$

Electrons in the incident beam can lose different amounts of energy in such encounters and typically a single electron will be brought to rest only after many such encounters. The X rays thus produced make up the continuous spectrum of Fig. 5-9b and are discrete photons whose wavelengths vary from λ_{min} to $\lambda \to \infty$, corresponding to the different energy losses in the individual encounters. The shortest wavelength photon would be emitted when an electron loses *all* its kinetic energy in *one* deceleration process; here $K' = 0$ so that $K = hc/\lambda_{min}$. Since K equals eV, the energy acquired by the electron in being accelerated through the potential difference V applied to the X-ray tube, we have

$$eV = hc/\lambda_{min}$$

or

$$\lambda_{min} = \frac{hc}{eV}. \tag{5-11}$$

Thus the minimum wavelength cutoff represents the complete conversion of the electron's kinetic energy to X-ray radiation. Equation 5-11 shows clearly that if $h \to 0$ then $\lambda_{min} \to 0$, which the prediction of classical theory. This shows that the very existence of a minimum wavelength is a quantum phenomenon.

The continuous X-ray radiation of Fig. 5-9b is often called *bremsstrahlung,* from the German *brems* (= braking, i.e., decelerating) + *strahlung* (= radiation). The bremsstrahlung process occurs not only in X-ray tubes but wherever fast electrons collide with matter, as in cosmic rays, in the Van Allen radiation belts which surround the earth, and in the stopping of electrons emerging from accelerators or radioactive nuclei. The bremsstrahlung process is sometimes regarded as the inverse of the Compton process: in the first an electron loses part of its (kinetic) energy to (create) a photon; in the second a photon loses part of its energy to an electron. However, a stronger analogy exists to the photoelectric effect, the bremsstrahlung process usually being considered to be an inverse photoelectric effect: in the photoelectric effect, a photon is absorbed, its energy and momentum going to an electron and its binding atom; in the bremsstrahlung process, a photon is created, its energy and momentum coming from the colliding electron and nucleus. In any case, we deal with the *creation* of photons in the bremsstrahlung process, rather than with their absorption or scattering by matter.

▶ **Example 5.** Determine Planck's constant h from the fact that the minimum X-ray wavelength produced by 40.0 kev electrons is 3.11×10^{-11} meter.

From Eq. 5-11, we have

$$h = \frac{eV \lambda_{min}}{c}$$

$$= \frac{(1.60 \times 10^{-19} \text{ coul})(4.00 \times 10^4 \text{ volts})(3.11 \times 10^{-11} \text{ m})}{(3.00 \times 10^8 \text{ m/sec})}$$

$$= 6.64 \times 10^{-34} \text{ joule-sec.}$$

This agrees well with the value of h deduced from the photoelectric effect and the Compton effect.

Measurement of V, λ_{min}, and c provides the present most accurate method for evaluating the ratio h/e. Bearden, Johnson, and Watts at the Johns Hopkins University found in 1951, using this procedure, $h/e = 1.37028 \times 10^{-15}$ joule-sec/coul. This ratio is combined with many other measured combinations of physical constants, the assembly of data being analyzed by elaborate statistical methods to find the "best" value for the various physical constants (see Ref. 5, e.g.). The best values change (but usually only within the a priori estimates of accuracy) and become increasingly precise as new experimental data and higher precision methods are used. ▶

5.6 Pair-Production and Pair-Annihilation

In addition to the photoelectric and Compton effects there is another process whereby photons lose their energy in interactions with matter, namely the process of *pair-production*. Pair production is also an excellent example of the conversion of radiant energy into rest mass energy as well as into kinetic energy. In this process a high energy photon loses all of its energy in an encounter with a nucleus (see Fig. 5-11), creating an electron and a positron (the *pair*) and endowing them with kinetic energies K_- and K_+. (A positron, discovered first by Anderson in the cosmic radiation, has the same mass and magnitude of charge as an electron, but its charge is positive rather than negative.) The energy taken by the recoil of the massive nucleus is negligible so that the relativistic energy balance can be written

$$h\nu = (m_0 c^2 + K_-) + (m_0 c^2 + K_+). \tag{5-12}$$

Here $2m_0 c^2$ is the (rest) energy needed to create the positron-electron pair, the positron and electron rest masses being equal. Although K_- and K_+ are

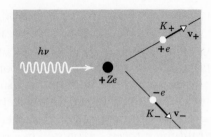

Figure 5-11. The pair-production process.

approximately equal, the positron has a somewhat larger kinetic energy than the electron. This is because of the Coulomb interaction of the pair with the (positively charged) nucleus which leads to an acceleration of the positron away from the nucleus and a deceleration of the electron.

In analyzing this process here we ignore the details of the interaction itself, considering only the situation before and after the interaction. Our guiding principles are the conservation of energy, conservation of momentum, and conservation of charge. From these (see Prob. 30) it follows that a photon cannot simply disappear in empty space, creating a pair as it vanishes, but that the presence of a massive particle is necessary to conserve both energy and momentum in the process. Charge is automatically conserved, the photon having no charge and the created pair of particles having no net charge. From Eq. 5-12 we see that the minimum (threshold) energy needed by a photon to create a pair is $2m_0c^2$ or 1.02 Mev; this corresponds to a wavelength of 0.012 Å. If the wavelength is shorter than this, corresponding to an energy greater than the threshold value, the photon endows the pair with kinetic energy as well as rest energy. The phenomenon is a high energy one, the photons being in the very short X-ray or the gamma-ray regions of the electromagnetic spectrum (these regions overlap; see Fig. 5-4). The photoelectric and Compton processes occur with the highest probability in the ultraviolet and long X-ray regions, respectively, whereas pair production occurs only at energies $\geq 2m_0c^2$ and primarily in the gamma ray or in the short X-ray region. Experiment confirms that the absorption of photons in its interaction with matter occurs principally by the photoelectric process at low energies, by the Compton effect at medium energies, and by pair production at high energies.

Electron-positron pairs are produced in nature by cosmic ray photons and in the laboratory by bremsstrahlung photons from particle accelerators. Other particle pairs, such as proton and antiproton (a particle with same mass and magnitude of charge as the proton, but with negative rather than positive charge), can be produced as well if the initiating photon has sufficient energy. Because the electron and positron have the smallest rest mass of known particles the threshold energy of their production is the smallest. Experiment verifies the quantum picture of the pair-production process. There is no satisfactory explanation whatever of this phenomenon in classical theory.

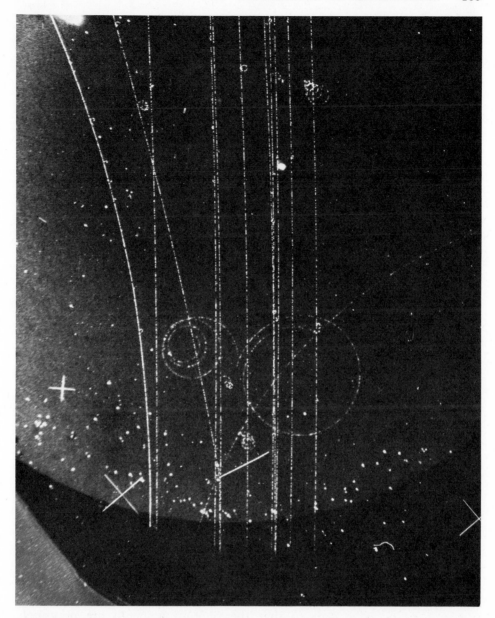

Figure 5-12. Electron pair-production, as seen in a bubble chamber. The electron and positron tracks are the two spirals meeting at the point where the production took place. The reader can determine which of the two spirals belongs to the positron by knowing that the long tracks are primarily deuterons (nuclei of the hydrogen isotope of atomic mass number 2) which are incident from the left. (Courtesy of C. R. Sun, State University of New York at Albany).

▶ **Example 6.** Analysis of a bubble chamber photograph (such as in Fig. 5-12) reveals the creation of an electron-positron pair as photons pass through a foil of matter. The electron and positron tracks have opposite curvatures in the uniform magnetic field of 0.20 weber/m^2, their radii being 2.5×10^{-2} m. What was the energy and the wavelength of the pair-producing photon?

The momentum of the electron (see Example 3 of Chapter 3) is given by

$$p\left(= mv = \frac{m_0 v}{\sqrt{1 - \beta^2}} \right) = eBr$$

$$= 1.6 \times 10^{-19} \times 2.0 \times 10^{-1} \times 2.5 \times 10^{-2} \text{ kg-m/sec}$$
$$= 8.0 \times 10^{-22} \text{ kg-m/sec.}$$

Its energy is given (see Chapter 3) by

$$E_-^2 = (m_0 c^2)^2 + (pc)^2.$$

But $m_0 c^2 = 0.51$ Mev and $pc = 8.0 \times 10^{-22} \times 3.0 \times 10^8$ joules
$$= 2.4 \times 10^{-13} \text{ joules} = 1.5 \text{ Mev}$$

so that $E_-^2 = (0.51 \text{ Mev})^2 + (1.5 \text{ Mev})^2$
and $E_- = 1.6$ Mev.

If the positron energy is $E_+ = K_+ + m_0 c^2$ and the electron energy is $E_- = K_- + m_0 c^2$, then (see Eq. 5-12) the energy of the photon was

$$h\nu = E_- + E_+ = 3.2 \text{ Mev}$$

where we have taken $E_+ = E_-$.
The photon's wavelength follows from

$$E = h\nu = hc/\lambda$$

or

$$\lambda = \frac{hc}{E} = \frac{6.6 \times 10^{-34} \times 3.0 \times 10^8}{3.2 \times 10^6 \times 1.6 \times 10^{-19}} \text{m} = 3.9 \times 10^{-13} \text{ m} = 0.0039 \text{ Å.} \blacktriangleleft$$

Closely related to the pair production is the inverse process called *pair annihilation*. An electron and a positron, which are essentially at rest near one another, unite and are annihilated. Matter disappears and in its place we get radiant energy. Since the initial momentum is zero and momentum must be conserved in the process, we cannot have only one photon created because a single photon cannot have zero momentum. The most probable process is the creation of two photons moving with equal momenta in opposite directions. Less probable, but possible, is the creation of three photons.

In the two photon process (see Fig. 5-13) momentum conservation gives $0 = \mathbf{p}_1 + \mathbf{p}_2$ or $\mathbf{p}_1 = -\mathbf{p}_2$ so that the photon momenta are oppositely directed but equal in magnitude. Hence, $p_1 = p_2$ or $h\nu_1/c = h\nu_2/c$ and $\nu_1 = \nu_2 = \nu$. Energy conservation then requires that $m_0 c^2 + m_0 c^2 = h\nu + h\nu$, the positron and electron having no initial kinetic energy and the photon energies being the same. Hence, $h\nu = m_0 c^2 = 0.51$ Mev, corresponding to a photon

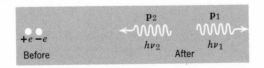

Figure 5-13. Pair-annihilation producing two photons.

wavelength of 0.024 Å. If the initial pair had some kinetic energy then the photon energy would exceed 0.51 Mev and its wavelength could be less than 0.024 Å.

Positrons are created in the pair production process. On passing through matter a positron loses energy in successive collisions until it combines with an electron to form a bound system called *positronium*. The positronium "atom" is short lived, decaying into photons within about 10^{-10} sec of its formation. The electron and positron presumably move about their common center of mass in a kind of "death dance" before mutual annihilation.

▶ **Example 7.** (*a*) Assume that Fig. 5-13 represents the annihilation process in a reference frame S, the electron-positron pair being at rest there and the two annihilation photons moving along the x-axis. Find the wavelength λ of these photons in terms of m_0, the rest mass of an electron or positron.

We saw above that $p_1 = p_2$ and $h\nu_1 = h\nu_2$. Each photon has the same energy, the same frequency, and the same wavelength. We can drop the subscripts then and from the relation $h\nu = m_0c^2$ and $p = E/c$ we obtain

$$p = \frac{E}{c} = \frac{h\nu}{c} = \frac{m_0c^2}{c} = m_0c.$$

But we also have the relation

$$p = h/\lambda$$

so that

$$\lambda = \frac{h}{p} = \frac{h}{m_0c}.$$

Hence, in the rest frame of the positronium atom each photon has the same wavelength, $\lambda = h/m_0c$.

(*b*) Now consider the same annihilation event to be observed in frame S', moving relative to S with a speed v to the left. What wavelength does this (moving) observer record for the annihilation photons?

Here, the pair has initial energy $2mc^2$, rather than merely the rest energy $2m_0c^2$, so that conservation of energy in the annihilation process gives us

$$2mc^2 = p'_1c + p'_2c.$$

Also, the pair now moves with speed v along the positive x'-axis so that its initial momentum is $2mv$, rather than zero as before. Conservation of momentum now gives us

$$2mv = p'_1 - p'_2,$$

the photons moving in opposite directions along the x'-axis. Let us combine these two expressions. We multiply the second by c and add it to the first, obtaining

$$p'_1 = m(c + v) = \frac{m_0(c + v)}{\sqrt{1 - v/c}} = m_0c\sqrt{\frac{c + v}{c - v}}.$$

But $p'_1 = h/\lambda'_1$, so that

$$\lambda'_1 = h/p'_1 = \frac{h}{m_0c}\sqrt{\frac{c - v}{c + v}} = \lambda\sqrt{\frac{c - v}{c + v}}.$$

Table 5-1

	Photoelectric Effect	Compton Effect	X-Ray Production	Pair Production	Pair Annihilation
Process Described	Photon absorbed and electron released from target	Photon scattered and electron recoils in target	Electron deflected in target and photon created	Photon absorbed and electron-positron pair created	Electron and positron combine and pair of photons created.
Principal Relation	$K_{max} = h\nu - w_0$	$\Delta\lambda = \dfrac{h}{m_0 c}(1 - \cos\phi)$	$h\nu = K - K'$	$h\nu = 2m_0 c^2 + (K_+ + K_-)$	$(m_0 c^2 + K_+) + (m_0 c^2 + K_-) = h\nu_1 + h\nu_2$
Presence of Other Particles	Electron must be bound to other particle—as an atom.	Electron can be free.	Electron deflected in encounter with heavy target nucleus.	Photon absorbed in encounter with heavy target nucleus.	Free electron and free positron, but two or more photons created.
Effective Energy and Wavelength Region	$h\nu > w_0$; energy a few electron volts and wavelength chiefly in ultraviolet.	Any energy, in principle. Most effective near 0.1 Mev and in long X-ray region.	$\lambda_{min} = hc/eV$ to $\lambda = \infty$. λ_{min} about 0.1 A and energy about 0.2 Mev or less.	Minimum $h\nu$ is $2m_0 c^2 =$ 1.02 Mev. Gamma ray region or short X-ray region.	Particles can be at rest. Minimum photoenergy 0.51 Mev and maximum wavelength 0.024 A.
Relative Probability	Less likely the higher the energy and frequency are beyond $h\nu_0 = w_0$. Goes to zero essentially at 1 Mev.	Less likely the higher the energy and frequency, but more likely than photoeffect beyond about 0.5 Mev.	Most effective near 1.5 to 2 times λ_{min}.	Probability increases as energy increases beyond minimum. More effective than Compton effect beyond about 5 Mev.	Two photon process more probable than three photon process.

In a similar manner, by subtracting the second equation from the first, we obtain

$$\lambda_2' = h/p_2' = \frac{h}{m_0 c} \sqrt{\frac{c+v}{c-v}} = \lambda \sqrt{\frac{c+v}{c-v}}.$$

The photons do *not* have the same wavelength, but are Doppler shifted from the wavelength λ they had in the rest frame of the source (the positronium atom). If an observer is situated on the x'-axis so that the source moves *toward* him, he will receive photon 1, having a frequency *higher* than the "rest" frequency. If an observer is situated on the x'-axis so that the source moves away from him, he will receive photon 2, having a frequency *lower* than the "rest" frequency.

Indeed, this example constitutes a derivation of the longitudinal Doppler effect alternative to the one presented in Chapter 2. ◀

In this chapter we have presented evidence for the particle nature of radiation. This evidence contradicts the classical picture of radiation as exhibiting only a wave nature. We have found that, as distinguished from its wavelike nature when it propagates, radiation is particlelike in its inter-action with matter. We have avoided the details of the interaction between matter and radiation, although a detailed theory (quantum electrodynamics) does now exist for all the processes we have discussed (pair-production, Compton scattering, photoelectric effect, etc.,). Such a theory explains the energy dependence of the probabilities of the various processes, their depend-ence on the atomic number of target materials, the angular distribution of scattered radiation and so forth. Instead, we have looked at the situation before and after the interaction, using the conservation principles of momen-tum and mass-energy to extract the principal features of the processes; see Table 5-1 for a summary. The breakdown of classical physics and the need for a new theory have emerged clearly. In the next chapter we continue to look at phenomena which help suggest how such a theory might be formu-lated.

QUESTIONS

1. In Fig. 5-2, why doesn't the photoelectric current rise vertically to its maximum (saturation) value when the applied potential difference is slightly more positive than $-V_s$?

2. Why is it that even for incident radiation that is monochromatic the photo-electrons are emitted with a spread of velocities?

3. Why are photoelectric measurements very sensitive to the nature of the photo-electric surface?

4. The existence of a cutoff frequency in the photoelectric effect is often regarded as the most potent objection to a wave theory. Explain.

5. Explain the statement that one could not detect faint starlight with his eyes if light were not corpuscular.

6. In the photoelectric equation $K = h\nu - w$ (Eq. 5-3), should we use $K = \frac{1}{2}m_0 v^2$ or $K = (m - m_0)c^2$? Does it really make any difference under ordinary circumstances?

7. Explain how a photoelectric cell can be used to open and close a supermarket door.

8. Can you use the device of letting $h \to 0$ to obtain classical results from quantum results in the case of the photoelectric effect? Explain.

9. Could one express the energy, E, of a photon in terms of its momentum and its mass? If so, give the relation.

10. In the photoelectric experiments, the current (number of electrons emitted per unit time) is proportional to the intensity of light. Show that the classical theory *or* the quantum theory could explain this result. Compare the explanations.

11. Assume that the emission of photons from a source of radiation is random in direction. Would you expect the intensity (or energy density) to vary inversely as the square of the distance from the source in the photon theory as it does in the wave theory?

12. Do the results of photoelectric experiments invalidate the results of Young's interference experiments? Does the interpretation of the results of one experiment invalidate the interpretation of the results of the other?

13. What is the direction of a Compton scattered electron with maximum kinetic energy compared to the direction of the incident monochromatic photon beam?

14. Why, in the Compton scattering picture, would you expect $\Delta\lambda$ to be independent of the materials of which the scatterer is composed?

15. Would you expect to observe the Compton effect more readily with scattering targets composed of atoms with high atomic number or those composed of atoms with low atomic number? Explain.

16. Light from distant stars is Compton scattered many times by free electrons in outer space before reaching us. This "shifts the light towards the red." How can this shift be distinguished from the Doppler "red shift" due to the motion of receding stars?

17. Explain the breadth of the Compton scattered lines (Fig. 5-6). That is, explain why we find scattered wavelengths other than λ and λ'.

18. Why don't we observe a Compton effect with visible light?

19. Would you expect a definite minimum wavelength in the emitted radiation for a given value of the energy of an electron incident on the target of an X-ray tube from the classical electromagnetic theory picture of the process?

20. Does a television tube emit X rays? Explain.

21. What effect(s) does decreasing the voltage across an X-ray tube have on the resulting X-ray spectrum?

22. Discuss the bremsstrahlung process as the inverse of the Compton process . . . of the photoelectric process.

23. Suppose you have a source of 200 kev X rays and you wish to demonstrate (*a*) the Compton effect and (*b*) the photoelectric effect. Would you use a target of low atomic number or high atomic number? Explain.

24. Describe several methods that can be used to determine experimentally the value of Planck's constant h.

25. From what factors would you expect to judge whether a photon will lose its energy in interactions with matter by the photoelectric process, the Compton process, or the pair production process?

26. Suppose that an electron-positron pair were in uniform motion relative to a laboratory observer at the moment of annihilation. Is it possible for them to create only a single photon? (Hint: use the principle of relativity.)

27. Could electron-positron annihilation occur with the creation of *one* photon if a nearby nucleus was available for recoil momentum?

28. Explain how pair annihilation with the creation of *three* photons is possible. Is it possible in principle to create even more than three photons in a single annihilation process?

29. What would be the inverse of the process in which two photons are created in electron-positron annihilation? Can it occur? Is it likely to occur?

30. Distinguish between Planck's formula $\Delta\varepsilon = h\nu$ (Eq. 4-9) and Einstein's formula $E = h\nu$ (Eq. 5-2)?

31. (*a*) Newton's light corpuscles were assumed to behave according to the laws of Newtonian mechanics. Is the photon concept a return to this idea of a light corpuscle? (*b*) The ether was invented as a medium in which light waves undulate. Does the photon concept eliminate the need for an ether?

PROBLEMS

1. (*a*) The energy required to remove an electron from sodium is 2.3 ev. Does sodium show a photoelectric effect for orange light, with $\lambda = 6800$ Å? (*b*) What is the cutoff wavelength for photoelectric emission from sodium? *Take hc = 12400 ev — Å.*

2. Light of a wavelength 2000 Å falls on an aluminum surface. In aluminum 4.2 ev are required to remove an electron. What is the kinetic energy of (*a*) the fastest and (*b*) the slowest emitted photoelectrons? (*c*) What is the stopping potential? (*d*) What is the cutoff wavelength for aluminum? (*e*) If the intensity of the incident light is 2.0 watt/m^2, what is the average number of photons per unit time per unit area that strike the surface?

3. The work function for a clean lithium surface is 2.3 ev. Make a rough plot of the stopping potential V_s versus the frequency of the incident light for such a surface. Indicate important features of the curve, such as slope and intercepts.

4. The stopping potential for photoelectrons emitted from a surface illuminated by light of wavelength $\lambda = 4910$ Å is 0.71 volts. When the incident wavelength is changed the stopping potential is found to be 1.43 volts. What is the new wavelength?

5. In a photoelectric experiment in which monochromatic light and a sodium photocathode are used, one finds a stopping potential of 1.85 volts for $\lambda = 3000$ Å and of 0.82 volts for $\lambda = 4000$ Å. From these data determine (*a*) a value for Planck's constant, (*b*) the work function of sodium in electron volts, and (*c*) the threshold wavelength for sodium.

6. You wish to pick a substance for a photocell operable with visible light. Which of the following will do (work function in parenthesis): tantalum (4.2 ev); tungsten (4.5 ev); aluminum (4.2 ev); barium (2.5 ev); lithium (2.3 ev)?

7. Consider light shining on a photographic plate. The light will be recorded if it dissociates a Ag Br molecule in the plate. The minimum energy to dissociate this molecule is of the order of 10^{-19} joule. (*a*) Show that there is a cutoff wavelength greater than which light will not be recorded. (*b*) Estimate the exposure time for incident light of power 10^{-2} watts to be recorded if classical theory applied. Does it?

8. The relativistic expression for kinetic energy should be used for the electron in the photoelectric effect when $v/c > 0.1$, if errors greater than about 1% are to be avoided. For photoelectrons ejected from an aluminum surface ($w_0 = 4.2$ ev) what is the smallest wavelength of an incident photon for which the classical expression may be used?

9. X rays with $\lambda = 0.71$ Å eject photoelectrons from a gold foil. The electrons form circular paths of radius r in a region of magnetic induction B. Experiment shows that $rB = 1.88 \times 10^{-4}$ tesla-meter. Find (a) the maximum kinetic energy of the photoelectrons and (b) the work done in removing the electron from the gold foil.

10. What is the frequency, wavelength, and momentum of a photon whose energy equals the rest energy of an electron?

11. (a) A spectral emission line, important in radioastronomy, has a wavelength of 21 cm. To what photon energy does this correspond? (b) The meter is defined to be 1,650,763.73 wavelengths of the orange radiation from Kr^{86}. How much energy is carried by one photon of this radiation?

12. Solar radiation falls on the earth at a rate of 1340 watts/m^2 on a surface normal to the incoming rays. Assuming an average wavelength of 5500 Å, how many photons/m^2-sec is this?

13. Under ideal conditions the normal human eye will record a visual sensation at 5500 Å if as few as 100 photons are absorbed per second. What power level does this correspond to?

14. A 100 watt sodium vapor lamp radiates uniformly in all directions. (a) At what distance from the lamp will the average density of photons be 10/cm^3? (b) What is the average density of photons 2.0 meters from the lamp? Assume the light to be monochromatic, with $\lambda = 5890$ Å.

15. In the photon picture of radiation, show that if beams of radiation of two different wavelengths are to have the same intensity (or energy density) then the numbers of the photons per unit cross-sectional area per second in the beams are in the same ratio as the wavelengths.

16. On the photon picture the intensity of a monochromatic electromagnetic wave is $Nh\nu$ where N is the number of photons crossing unit area per unit time. What is the radiation pressure? (See *Physics* Part II, Sec. 40-2.) For a given intensity does the pressure depend on the wavelength of the radiation?

★**17.** (a) Show that a free electron cannot absorb a photon and conserve both energy and momentum in the process. Hence, the photoelectric process requires a bound electron. (*Hint.* Assume energy conservation and show that momentum is not conserved.) (b) In the Compton effect, however, the electron *can* be free. Explain.

★**18.** Eliminate v and θ from Eqs. 5-6, 5-7, 5-8 and derive Eq. 5-9, $\Delta\lambda = (h/m_0c)(1 - \cos\phi)$. (*Hint.* Simplify the notation, by letting $1/\sqrt{1 - v^2/c^2} = \gamma$ for example; divide Eq. 5-6 by m_0c^2 and Eq. 5-7 and 5-8 by m_0c; and use the relation $\sin^2 + \cos^2 = 1$.)

19. Derive the relation

$$\cot\frac{\phi}{2} = \left(1 + \frac{h\nu}{m_0c^2}\right)\tan\theta$$

between the direction of motion of the scattered photon and the recoil electron in the Compton effect. Start from the conservation relations, Eqs. 5-6 to 5-8, or from equations (such as the Compton result) derived from them.

20. Derive a relation between the kinetic energy K of the recoil electron and the energy E of the incident photon in the Compton effect, starting from the conservation relation Eqs. 5-6 to 5-8. One form of the relation is

$$\frac{K}{E} = \frac{(2h\nu/m_0c^2)\sin^2\phi/2}{1 + (2h\nu/m_0c^2)\sin^2\phi/2}.$$

(*Hint.* See Example 4 and note that $(1 - \cos\phi) = 2\sin^2\phi/2$.)

21. What is the maximum kinetic energy of the Compton scattered electrons from a sheet of copper struck by a monochromatic photon beam in which the incident photons each have a momentum of 0.88 Mev/c?

22. What is the maximum possible kinetic energy of a recoiling Compton electron in terms of the incident photon energy $h\nu$ and the electron's rest energy m_0c^2? (*Hint.* See Prob. 20).

23. Photons of wavelength 0.024 Å are incident on free electrons. (*a*) Find the wavelength of a photon which is scattered 30° from the incident direction and the kinetic energy imparted to the recoil electron. (*b*) Do the same if the scattering angle is 120°. (*Hint.* See Example 4).

24. An X-ray photon of initial energy 1.0×10^5 ev traveling in the $+x$-direction is incident on a free electron at rest. The photon is scattered at right angles into the $+y$-direction. Find the components of momentum of the recoiling electron.

25. (*a*) Show that $\Delta E/E$, the fractional change in photon energy in the Compton effect, equals $(h\nu'/m_0c^2)(1 - \cos\phi)$. (*b*) Plot $\Delta E/E$ versus ϕ and interpret the curve physically.

26. Determine the maximum wavelength shift in the Compton scattering of photons from *protons*.

27. (*a*) Show that the short wavelength cutoff in the X-ray continuous spectrum is given by $\lambda_{\min} = 12.4$ Å/V, where V is applied voltage in kilovolts. (*b*) If the voltage across an X-ray tube is 186 kv what is $\lambda_{\min}$?

28. (*a*) What is the minimum voltage across an X-ray tube that will produce an X ray having the Compton wavelength? ... a wavelength of 1 Å? (*b*) What is the minimum voltage needed across an X-ray tube if the subsequent bremsstrahlung radiation is to be capable of pair production?

29. What is the threshold energy for a photon to produce a proton-antiproton pair in the vicinity of a heavy nucleus.

★**30.** A gamma ray creates an electron-positron pair. Show directly that, without the presence of a third body to take up some of the momentum, energy and momentum cannot both be conserved. (*Hint.* Set the energies equal and show that this leads to unequal momenta before and after the interaction.)

31. A gamma ray can produce an electron-positron pair in the neighborhood of an electron at rest as well as near a nucleus. Show that in this case the threshold energy is $4m_0c^2$. (*Hint.* Do not ignore the recoil of the original electron but assume that all three particles move off together.)

32. A particular pair is produced such that the positron is at rest and the electron has a kinetic energy of 1.0 Mev moving in the direction of flight of the pair producing photon. (*a*) Neglecting the energy transferred to the nucleus of the nearby atom, find the energy of the incident photon. (*b*) What percentage of the photon's momentum is transferred to the nucleus?

33. Assume that an electron-positron pair is formed by a photon having the threshold energy for the process. (*a*) Calculate the momentum transferred to the nucleus in the process. (*b*) Assume the nucleus to be that of a lead atom (atomic weight 207.2 a.m.u.) and compute the kinetic energy of the recoil nucleus. Are we justified in neglecting this energy compared to the threshold energy assumed above?

34. An electron-positron pair annihilate *in flight* producing two photons of wavelength λ_1 and λ_2. Write down the equations (do *not* solve them) from which one may determine the angle between the two created photons in terms of their wavelengths λ_1 and λ_2 (see Fig. 5-14).

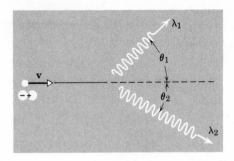

Figure 5-14. Illustrating Problem 34.

★**35.** In frame S a photon of 1.02 Mev has just enough energy to create a pair. Frame S' is an inertial frame moving at a speed $\frac{3}{5} c$ relative to frame S as shown in the figure. To an observer in the S' frame, the photon will be Doppler shifted (see Fig. 5-15). (*a*) Calculate, from the Doppler shift in wavelength, the energy of the photon as measured by an observer in S'. (*b*) The energy of the photon in S' is less than the threshold energy of 1.02 Mev. But pair creation is a physical process. It either happens or it doesn't. The fact of its occurrence cannot depend on which frame you choose to view it. Explain this apparent paradox. (*Hint.* What role does the third body play? Assume it is at rest in frame S.)

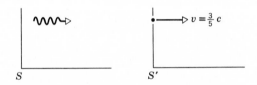

Figure 5-15. Illustrating Problem 35.

REFERENCES

1. Albert Einstein, "On a Heuristic Viewpoint Concerning the Production and Transformation of Light," *Annalen der Physik,* **17,** 132 (1905); translated into English, in *Great Experiments in Physics,* edited by Morris H. Shamos, Holt-Dryden 1959.

2. For an historical account of Compton's researches read A. H. Compton, "The Scattering of X-Rays as Particles," *Amer. J. Phys.* 817 (Dec. 1961).

3. G. Gamow, *Mr. Tompkins in Wonderland,* Cambridge University Press, Cambridge, 1957.

4. A. H. Compton, "A Quantum Theory of the Scattering of X-Rays by Light Elements," *Phys. Rev.* **21,** 483 (1923). Extracted in *Great Experiments in Physics,* edited by Morris H. Shamos, Holt-Dryden, 1959.

5. J. H. Sanders, "Fundamental Constants: Their Relationship and Measurement," *Physics Education,* 177 (July 1968).

CHAPTER 6

The Wave Nature of Matter
and the Uncertainty Principle

6.1 Matter Waves

Maurice de Broglie was a French experimental physicist who, from the outset, had supported Compton's view of the particle nature of radiation. His experiments and discussions impressed his brother Louis so much with the philosophic problems of physics at the time that Louis changed his career from history to physics. In his doctoral thesis, presented in 1924 to the Faculty of Science at the University of Paris, Louis de Broglie proposed the existence of matter waves (see Ref. 1). The thoroughness and originality of his thesis were recognized at once, but because of the apparent lack of experimental evidence de Broglie's ideas were not considered to have any physical reality. It was Albert Einstein, whose attention was drawn to de Broglie's ideas by Paul Langevin, who recognized their importance and validity and in turn called them to the attention of other physicists. Five years later de Broglie became the first physicist ever to receive a Nobel prize for his doctoral thesis, his ideas by then having been dramatically confirmed by experiment.

De Broglie's hypothesis was that the apparently dual behavior of radiation as waves and as particle applied equally well to matter. Just as a quantum of radiation has a wave associated with it that governs its motion, so a quantity of matter will have a corresponding matter wave that governs its motion. Since the observable universe is composed entirely of matter and radiation, de Broglie's suggestions are consistent with a statement of the symmetry in nature. Indeed, de Broglie proposed that the wave aspects of matter were related quantitatively to the particle aspects in exactly the same way that we found for radiation, namely $E = h\nu$ and $p = h/\lambda$.

That is, for matter *and* for radiation alike the total energy E of the entity is related to the frequency ν of the wave associated with its motion by the equation

$$E = h\nu \tag{6-1a}$$

and the momentum p of the entity is related to the wavelength λ of the associated wave by the equation

$$p = h/\lambda. \tag{6-1b}$$

Here the particle concepts, energy E and momentum p, are connected to the wave concepts, frequency ν and wavelength λ. Radiation corresponds to particles of zero rest mass (moving at speed c), whereas matter corresponds to particles of finite rest mass (moving at speeds less than c), but the same general transformation properties apply to both entities. We saw in Chapter 5 that radiation, which earlier was regarded as wavelike, exhibited particlelike properties as well. De Broglie suggested that matter, which had been regarded as particlelike, exhibits wavelike properties also. Equation 6-1b in the following form, is called the de Broglie relation

$$\lambda = \frac{h}{p}. \tag{6-2}$$

It predicts the wavelength λ of a *matter wave*, associated with the motion of matter having a momentum p.

▶ **Example 1.** (*a*) What is the de Broglie wavelength of a baseball moving at a speed $v = 10$ m/sec? Assume $m = 1.0$ kg. From Eq. 6-2

$$\lambda = \frac{h}{p} = \frac{h}{mv} = \frac{6.6 \times 10^{-34} \text{ joule-sec}}{(1.0)(10) \text{ kg m/sec}}$$

$$= 6.6 \times 10^{-35} \text{ m} = 6.6 \times 10^{-25} \text{ Å}.$$

(*b*) What is the de Broglie wavelength of an electron whose kinetic energy is 100 ev? Here

$$\lambda = \frac{h}{p} = \frac{h}{\sqrt{2mK}}$$

$$= \frac{6.6 \times 10^{-34} \text{ joule-sec}}{(2 \times 9.1 \times 10^{-31} \text{ kg} \times 100 \text{ ev} \times 1.6 \times 10^{-19} \text{ joule/ev})^{1/2}}$$

$$= \frac{6.6 \times 10^{-34} \text{ joule-sec}}{5.4 \times 10^{-24} \text{ kg m/sec}} = 1.2 \times 10^{-10} \text{ m} = 1.2 \text{ Å}. \quad ◀$$

The wave nature of light propagation is not revealed by experiments in geometrical optics, for the obstacles or apertures used there are very large compared to the wavelength of light. If a represents a characteristic dimension of such an aperture or obstacle (e.g., width of a slit) we say that $\lambda/a \to 0$ in the region of geometric optics. Geometrical optics is characterized by ray propagation, which is similar to the trajectory motion of particles. When the dimension a becomes comparable to λ (that is, $\lambda \simeq a$) or smaller than λ (that is, $\lambda > a$), then we are in the region of physical optics. In this case interference and diffraction effects are easily observed and wavelengths can be measured readily. To observe the wavelike aspects of the motion of matter, therefore, we need suitably small apertures or obstacles. One of the smallest sizes available at the time of de Broglie was that of an atom, or the spacing between adjacent planes of atoms in a solid, where $a \simeq 1$ Å. Clearly then, for the baseball in Example 1 we cannot expect to measure the de Broglie wavelength (here $\lambda/a \simeq 10^{-25}$); indeed we cannot detect *any* evidence of wavelike motion. For material objects of very much smaller mass, however, such as an electron ($\lambda \simeq a$ in Example 1), we might detect an associated wave because the smaller momentum corresponds to a larger wavelength.

It was Elsasser who pointed out, in 1926, that the wave nature of matter might be tested in the same way that the wave nature of X rays was first tested, namely by causing a beam of electrons of appropriate energy to fall on a crystalline solid. The atoms of the crystal serve as a three-dimensional array of diffracting centers for the electron "wave." We should look for strong diffracted peaks, then, in certain characteristic directions, just as for X-ray diffraction (cf. *Physics,* Part II Sec. 45.5). This idea was confirmed by C. J. Davisson and L. H. Germer in the United States (see Ref. 2) and by G. P. Thomson in Scotland (see Ref. 3).

Figure 6-1 shows schematically the apparatus of Davisson and Germer.* Electrons from a heated filament are accelerated through a potential difference V and emerge from the "electron gun" G with kinetic energy eV. This electron beam falls at normal incidence on a single crystal of nickel at C. The detector D is set at a particular angle ϕ and readings of the intensity of the "reflected" beam are taken at various values of the accelerating potential V. Figure 6-2, for example, shows that a strong beam is detected at $\phi = 50°$ for $V = 54$ volts. All such strong "reflected" beams can be accounted for by assuming that the electrons have a wavelength given by $\lambda = h/p$ and that "Bragg reflections" occur from certain families of atomic planes precisely

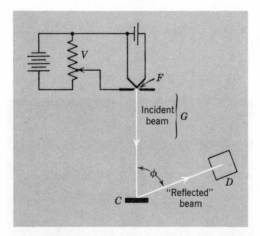

Figure 6-1. The apparatus of Davisson and Germer. Electrons from filament F are accelerated by a variable potential difference V. After "reflection" from crystal C they are collected by detector D.

*An explosion of a liquid-air bottle opened the vacuum system to the air and oxidized the target. In heating the target afterwards to get it clean, Davisson and Germer recrystallized the polycrystalline target into a few large crystals which accidently created the conditions that, according to K. K. Darrow, "blew open the gate to the discovery of electron waves." See Ref. 1.

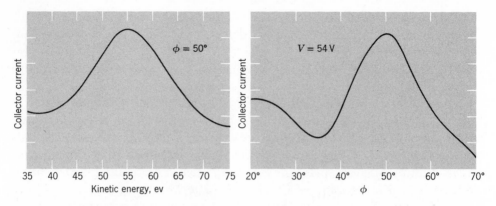

Figure 6-2. (*a*) The collector current in detector *D* in Fig. 6-1 as a function of the kinetic energy of the incident electrons, showing diffraction maximum. The angle φ in Fig. 6-1 is adjusted to 50°. If an appreciably smaller or larger value is used, the diffraction maximum disappears. (*b*) The current as a function of detector angle φ for fixed value 54 ev electron kinetic energy.

as described for X rays (*Physics,* Part II, Sec. 45-5). The observations *cannot* be accounted for on the basis of particle motion, but *can* be explained in terms of interferences. In classical terms, particles cannot interfere but waves can.

Figure 6-3 shows such a Bragg reflection, obeying the Bragg relationship

$$m\lambda = 2d \sin \theta \qquad m = 1, 2, 3, \ldots \tag{6-3}$$

For the conditions of Fig. 6-3 the effective interplaner spacing *d* can be shown by X-ray analysis to be 0.91 Å. Because of symmetry about the perpendicular to the lattice planes, $\phi + 2\theta = 180°$. Since $\phi = 50°$, it follows that $\theta = 65°$.

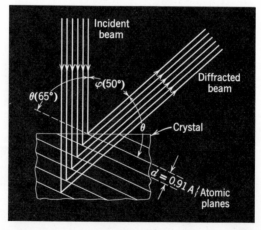

Figure 6-3. The strong diffracted beam at $\phi = 50°$ and $V = 54$ volts arises from wave like "reflection" from the family of atomic planes shown, for $d = 0.91$ Å. The Bragg angle θ is 65°. For simplicity, refraction of the diffracted wave as it leaves the surface is not shown.

The wavelength calculated from Eq. 6-3, assuming $m = 1$, is

$$\lambda = 2d \sin \theta = 2(0.91 \text{ Å}) \sin 65° = 1.65\text{Å}.$$

The de Broglie wavelength for 54 ev electrons, calculated from Eq. 6-2, is

$$\lambda = h/p = 6.6 \times 10^{-34} \text{ joule-sec}/4.0 \times 10^{-24} \text{ kg-m/sec} = 1.65 \text{ Å}.$$

This is quantitative confirmation of de Broglie's equation for λ in terms of p. The choice of $m = 1$ above is justified by the fact that if $m = 2$ (or more) then other reflection peaks for different angles ϕ would have appeared, but none were observed (see Prob. 10). The breadth of the observed peak in Fig. 6-2 is easily understood, also, for low energy electrons cannot penetrate deeply into the crystal so that only a small number of atomic planes contribute to the diffracted wave. Hence, the diffraction maximum is not sharp. Indeed, all the experimental results were in excellent qualitative and quantitative agreement with the de Broglie prediction, and provided convincing evidence that matter moves in agreement with the laws of wave motion.

In 1927, George P. Thomson showed that electron beams are diffracted in passing through thin films and independently confirmed the de Broglie relation $\lambda = h/mv$ in detail. Whereas the Davisson-Germer experiment is like Laue's in X-ray diffraction (reflection of specific wavelengths in a continuous spectrum from the regular array of atomic planes in a large single crystal), Thomson's experiment is similar to the Debye-Hull-Scherrer method of powder diffraction of X rays (transmission of a fixed wavelength through an aggregate of very small crystals oriented at random). Thomson used higher energy electrons, which are much more penetrating, so that many hundred atomic planes contribute to the diffracted wave. The resulting diffraction pattern consists of sharp lines. In Fig. 6-4 we show, for comparison, an X-ray diffraction pattern and an electron diffraction pattern from polycrystalline substances (substances in which a large number of microscopic crystals are oriented at random).

It is of interest that J. J. Thomson, who in 1897 discovered the electron (which he characterized as a particle with a definite charge-to-mass ratio) and was awarded the Nobel prize in 1906, was the father of G. P. Thomson, who in 1927 experimentally discovered electron diffraction and was awarded the Nobel prize (with Davisson) in 1937. Max Jammer (p. 254 of Ref. 4) says of this "one may feel inclined to say that Thomson, the father, was awarded the Nobel prize for having shown that the electron is a particle, and Thomson, the son, for having shown that the electron is a wave."

Not only electrons but *all material objects*, charged or uncharged, show wavelike characteristics in their motion under the conditions of physical optics. For example, Estermann, Stern and Frisch performed quantitative experiments on the diffraction of molecular beams of hydrogen and atomic beams of helium from a lithium fluoride crystal; and Fermi, Marshall, and Zinn showed interference and diffraction phenomena for slow neutrons. In Fig. 6-5 we show a Laue diffraction pattern for neutrons. Indeed, even an interferometer operating with electron beams has been constructed (Ref. 5). The existence of matter waves is well established.

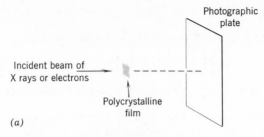

Figure 6-4. (*a*) The experimental arrangement for Debye-Scherrer diffraction of X rays or electrons by a polycrystalline material.

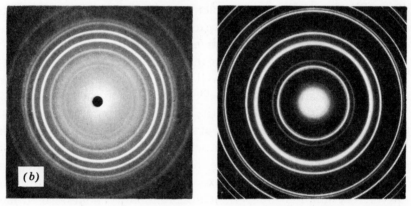

(*b*) *Left:* Debye-Scherrer pattern of *X-ray diffraction* by zirconium oxide crystals. *Right:* Debye-Scherrer pattern of *electron diffraction* by gold crystals. From U. Fano and L. Fano, *Basic Physics of Atoms to Molecules,* John Wiley and Sons, 1959.

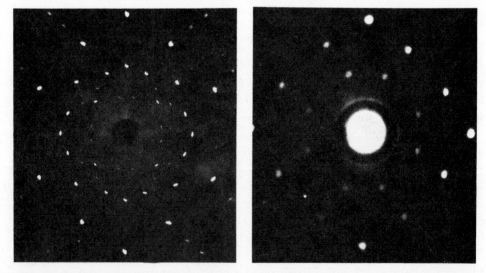

Figure 6-5. *Left:* Laue pattern of X-ray diffraction by a single sodium chloride crystal. (Courtesy of W. Arrington and J. L. Katz, X-Ray Laboratory, Rensselaer Polytechnic Institute). *Right:* Laue pattern of diffraction of neutrons from a nuclear reactor by a single sodium chloride crystal. (From Blackwood et al., *Outline of Atomic Physics,* John Wiley and Sons, 1955.)

It is instructive to note that we had to go to long de Broglie wavelengths to find experimental evidence for the wave nature of matter. That is, we used particles of low mass and speed to bring $\lambda(= h/mv)$ into the range of measurable diffraction. Ordinary (macroscopic) matter has such short corresponding de Broglie wavelengths, because its momentum is high, that its wave aspects are practically undetectable. The particle aspects are dominant. Similarly, we had to go to very short wavelengths to find experimental evidence for the particle nature of radiation. It is in the X-ray and gamma-ray region where the corpuscular aspects of radiation stand out experimentally. In the long wavelength region classical wave theory is completely adequate to explain the observations. All this suggests that just as a quantum theory of radiation emerges from the experiments of earlier chapters so a wave mechanics of particles is emerging from experiments discussed here.

Once again we see the central role played by Planck's constant h. If h were zero then, in $\lambda = h/mv$, we obtain $\lambda = 0$. Matter would always have a wavelength smaller than *any* characteristic dimension and diffraction could never be observed; this is the classical situation. Furthermore, we see that it is the smallness of h that obscures the existence of matter waves experimentally in the macroscopic world for we must have very small momenta to obtain measurable wavelengths. For ordinary macroscopic matter m is large and λ is so small as to be beyond the range of experimental measurement, so that classical mechanics is supreme. But in the microscopic world, de Broglie wavelengths are comparable to characteristic dimensions of systems of interest (such as atoms), so that the wave properties of matter in motion are experimentally observable.

◆ **Example 2.** In the experiments with helium atoms referred to above, a beam of atoms of nearly uniform speed of 1.635×10^5 cm/sec was obtained by allowing helium gas to escape through a small hole in its enclosing vessel into an evacuated chamber and then through narrow slits in parallel rotating circular disks of small separation (a mechanical velocity selector.) In addition to a regularly reflected beam, a strongly diffracted beam of helium atoms was observed to emerge from the lithium fluoride crystal surface upon which the atoms were incident. The diffracted beam was detected with a highly sensitive pressure gauge. The usual crystal diffraction analysis indicated a wavelength of 0.600×10^{-8} cm. What was the de Broglie wavelength?

The mass of a helium atom is

$$m = \frac{M}{N_0} = \frac{4.00 \times 10^{-3} \text{ kg/mole}}{6.02 \times 10^{23} \text{ atoms/mole}} = 6.65 \times 10^{-27} \text{ kg.}$$

According to the de Broglie equation the wavelength then is

$$\lambda = \frac{h}{mv} = \frac{6.63 \times 10^{-34} \text{ joule-sec}}{6.65 \times 10^{-27} \text{ kg} \times 1.635 \times 10^3 \text{ m/sec}} = 0.609 \times 10^{-10} \text{ m}$$

$$= 0.609 \times 10^{-8} \text{ cm.}$$

This result, 1.5% greater than the value measured by crystal diffraction, was well within the limits of error of the experiment.

Such experiments are very difficult to perform in any case, but especially since the intensities obtainable in atomic beams are quite low. Neutron diffraction experiments, using crystals of known lattice spacing, give confirmation of the existence of matter waves and precise confirmation of de Broglie's equation. The precision is due to the fact that the supply of neutrons from nuclear reactors is copious. Indeed, neutron diffraction is now an important method of studying crystal structure. Certain crystals, such as hydrogenous organic ones, are particularly well suited to neutron diffraction analysis, since neutrons are strongly scattered by hydrogen atoms whereas X rays are very weakly scattered by them. X rays interact chiefly with electrons in the atom, and electrons interact with the nuclear charge of the atom as well as the atomic electrons (electromagnetic forces) so that their interaction with hydrogen atoms is weak (the charge is small). But neutrons (which have no charge) interact principally with the nucleus of the atom (nuclear forces), and the interaction is strong. ◀

6.2 Electron Optics and the Electron Microscope

We can draw an analogy between geometrical optics and classical mechanics by means of a simple example. Consider a beam of electrons moving from a region in which the electrostatic potential is V_1 to a region in which the potential is V_2. Let $V_2 > V_1$ and imagine that V changes continuously in a narrow transition strip at the interface between the regions (see Fig. 6-6). In the transition strip a force per unit charge acts on each electron given by $-dV/dy$ (see the figure) so that electrons are accelerated in the negative y-direction, the x-component of velocity remaining unchanged, however. If $\mathbf{p}_1$ is the electron momentum in region 1 and $\mathbf{p}_2$ in region 2, then from the constancy of p_x we have

$$p_1 \sin \theta_1 = p_2 \sin \theta_2$$

where θ_1 and θ_2 are the angles made by the beams with the normal to the interface. Now, this invites a comparison with the refraction of light rays passing from a medium of refractive index n_1 into one of refractive index n_2, where $n_2 > n_1$. In such a case the rays are bent toward the normal, following Snell's law of refraction.

$$n_1 \sin \theta_1 = n_2 \sin \theta_2.$$

If we wish to draw an analogy we would say that the corresponding "refractive index" for an electron is proportional to the electron momentum; that is, $n \propto p$.

If the total energy of an electron in an electrostatic field is E, then $E = K + U = p^2/2m - eV$ for nonrelativistic speeds. Hence, $p = \sqrt{2m(E + eV)}$, where E is the constant (conserved) total energy and eV is the electrostatic potential energy. If the electron refractive index is proportional to p, we see that the "refractive index" varies with position y in the inhomogeneous transition strip because V varies there. Also, the regions have different "refractive indices" for electrons of different total energy E, in analogy to the optical phenomenon of dispersion wherein n varies with the frequency of light.

The motion of electrons in an electron microscope can be described in terms of an equivalent refractive index, therefore, just as the paths of light rays are described in an optical microscope. But, where does the de Broglie wavelength of electrons enter this analogy? Let's pursue

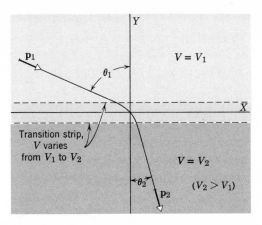

Figure 6-6. Illustrating the "refraction" of an electron beam as it passes through a region of varying potential.

the analogy a bit farther. In geometrical optics we conclude (see Chap. 41 of *Physics,* Part II) that the wavelengths change inversely as the refractive indices across an interface. That is, $\lambda_2/\lambda_1 = n_1/n_2$ or $\lambda \propto 1/n$. The corresponding electron refractive index is proportional to p however, so we expect $\lambda \propto 1/p$ or $\lambda = \text{constant}/p$. If we call the constant h, we have $\lambda = h/p$. Experiment determines h to be Planck's constant. Hence, our electron wavelength becomes

$$\lambda = h/\sqrt{2m(E + eV)}. \tag{6-4a}$$

Now we have usually taken the zero of potential ($V = 0$) to be where the electron is at rest ($K = 0$), so that the conserved total energy $E = K + eV$ equals zero. With $E = 0$, then, we obtain

$$\lambda = \frac{h}{\sqrt{2meV}} = \frac{12.3}{\sqrt{V}} \text{ Å} \tag{6-4b}$$

as the wavelength in angstroms when V is in volts. Here V is the accelerating potential, numerically equal to the kinetic energy of the electrons in electron volts.

What connection does this matter wavelength have with the electron microscope? Recall from our study of physical optics of light that the wavelength of the light used in a microscope sets a limit to the resolving power of the instrument. Similarly, the wave nature of electrons (responsible for a physical optics of electrons) reveals itself by the fact that the de Broglie wavelength sets a theoretical limit to the resolving power of an electron microscope. For a circular lens we can resolve two objects if their angular separation is greater than θ_R, where $\theta_R = 1.22 \lambda/d$. Here d is the diameter of the circular lens, λ is the wavelength of light used, and the factor 1.22 comes from the circular geometry (see *Physics,* Part II, Sec. 44-5). The important factor to note here is λ, the wavelength; the smaller the wavelength used, the closer together two objects can be and still be resolved. This accounts for the use of shorter wavelengths in the visible spectrum with optical microscopes, blue light giving better resolution than red light, for example. It suggests that X rays, having a wavelength smaller than the visible by a factor of 1000, could give enormously better resolution; the problem is to build an X-ray microscope, a task not yet achieved. The refractive index of matter for X rays is practically unity, making appropriate lens systems impractical. However, we have seen that electrons have matter waves whose wavelengths can be in the X-ray wavelength region. An electron beam can be deflected by electric and magnetic fields so that an electron lens system can be devised (see Fig. 6-7). Hence, an electron microscope can be built.

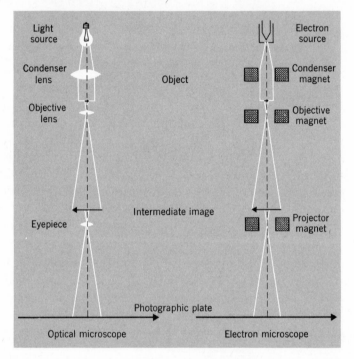

Figure 6-7. Comparison of an optical microscope with an electron microscope using a magnetic lens system. The electron microscope provides an ultimate magnification about 100 times greater and a resolution about 1000 times greater than that of the optical microscope. (The eyepiece of the optical microscope is adjusted to project a real image on the photographic plate, rather than providing a virtual image as would be required for direct visual observation.)

The small de Broglie wavelengths of electron beams give a resolving power much greater than that attainable with ordinary light. And, for the same wavelength, the electron kinetic energy is much smaller than the energy of the corresponding photon (see Prob. 15) so that electrons will scatter elastically off specimens whose structure would be altered by photons of the same wavelength. Furthermore, one can vary the wavelength conveniently in an electron microscope by changing the accelerating voltage V (see Eq. 6-4b). Apart from their use in a microscope, electron diffraction techniques may have other advantages over X-ray diffraction ones for many materials because much higher intensities are available for electron exposure than for X-ray exposure.

▶ **Example 3.** The smallest linear separation resolvable in principle by a microscope is about equal to the wavelength used. An electron microscope uses electrons of 100 Kev energy. Determine the theoretical limit for this electron microscope.

From Eq. 6-4*b* we have

$$\lambda = \frac{h}{\sqrt{2meV}} = \frac{12.3}{\sqrt{V}} \text{ Å} = \frac{12.3}{\sqrt{10^5}} \text{ Å} = 0.039 \text{ Å}.$$

This is a nonrelativistic result. If we were to use the relativistic expression for p in $\lambda = h/p$ we would obtain a somewhat smaller wavelength (see Prob. 12).

In practice, however, the best resolution obtained has been 5 Å with electron beams incident on suitable specimens. Practical limitations are imposed by spherical aberration in electron lens systems and loss of contrast under the best conditions for resolution.

In Fig. 6-8 we show an electron micrograph of a cancer cell.

Figure 6-8. A three-dimensional electron micrograph of a cancer cell, taken by a stereoscopic electron microscope. A conventional electron microscope can focus only on a plane at a fixed depth in the object studied, so it produces a two-dimensional image. This microscope has a scanning focal plane that produces three-dimensional images seen on a TV screen; it has a magnification of 100,000 with a resolution up to 150 angstroms. From "Scientific Research" Vol. 3, No. 4, 1968. (See Ref. 6 for articles on electron microscopy.)

Electron wavelengths can be very much shorter than the wavelength of X rays produced in an X-ray tube. The Stanford electron linear accelerator, for example, provides electrons of wavelength near 10^{-14} cm. The accelerator is used in a manner analogous to a microscope, in the sense that it is used to study the size and shape of nuclei and nucleons by diffracting the electrons from such nuclei or nucleons. By observing the diffraction patterns one can then infer the shape of the diffracting targets. ◀

6.3 The Wave-Particle Duality

In our studies in classical physics we have seen that energy is transported either by waves or by particles. Water waves carry energy over the water surface and bullets transfer energy from gun to target, for example. From such experiences we built a wave model for certain macroscopic phenomena and a particle model for other macroscopic phenomena and quite naturally extended these models into visually less accessible regions. Thus we explained sound propagation in terms of a wave model and pressures of gases in terms of a particle model (kinetic theory). Our successes conditioned us to expect that entities either are particles or are waves. Indeed, these successes extended into the early twentieth century with applications of Maxwell's wave theory to radiation and the discovery of elementary particles of matter, such as the neutron and positron.

Hence, we were quite unprepared to find that to understand radiation we need to invoke a particle model in some situations, as in the Compton effect, and a wave model in other situations, as in the diffraction of X rays. Perhaps more striking is the fact that this same wave particle duality applies to matter as well as to radiation. The charge-to-mass ratio of the electron and its ionization trail in matter (a sequence of localized collisions) suggest a particle model, but electron diffraction suggests a wave model. We have been compelled to use both models for the same entity. It is important to note, however, that in any given measurement only one model applies—we do not use both models under the same circumstances. When we detect the entity by some kind of interaction, it acts like a particle in the sense that it is localized; when moving it acts like a wave in the sense that interference phenomena are observed, and, of course, a wave is extended, not localized.

Neils Bohr summarized the situation in his principle of *complementarity*. The wave and particle models are complementary: if a measurement proves the wave character of radiation or matter, then it is impossible to prove the particle character in the same measurement, and conversely. Which model we use is determined by the nature of the measurement. Furthermore, our understanding of radiation, or of matter, is incomplete unless we take into account measurements which reveal the wave aspects and also those that reveal the particle aspects. Hence, radiation and matter are not simply waves nor simply particles. A more general and, to the classical mind, a more complicated model is needed to describe their behavior, even though in extreme situations a simple wave model or a simple particle model may apply.

The link between wave model and particle model is provided by a probabilistic interpretation of the wave particle duality. In the case of radiation it was Einstein who united the wave and particle theories; subsequently Max Born applied a similar argument to unite wave and particle theories of matter.

On the wave picture the intensity of radiation I, is proportional to $\overline{\mathcal{E}^2}$ where $\overline{\mathcal{E}^2}$ is the average value over one cycle of the square of the electric field strength of the wave, ('Physics' Part II, Sec. 39-6). (I is the average value of the Poynting vector and we use the symbol $\mathcal{E}$ instead of E for electric field to avoid confusion with the total energy E.) On the photon, or particle, picture the intensity is written as $I = Nh\nu$ where N is the average number of photons per unit time crossing unit area perpendicular to the direction of propagation. It was Einstein who suggested that $\overline{\mathcal{E}^2}$, which in electromagnetic theory is proportional to the radiant energy in a unit volume, could be interpreted as a measure of the average number of photons per unit volume.

Recall that Einstein introduced a granularity to radiation, abandoning the continuum interpretation of Maxwell. This leads to a statistical view of intensity. In this view, a point source of radiation emits photons randomly in all directions. The average number of photons crossing a unit area will decrease with the distance the area is from the source. This is because the photons spread out over a larger cross sectional area the farther they are from the source. Since this area is proportional to the distance squared, we obtain, on the average, an inverse square law of intensity just as we did on the wave picture. In the wave picture we imagined spherical waves to spread out from the source, the intensity dropping inversely as the square of the distance from the source. Here, these waves, whose strength can be measured by $\overline{\mathcal{E}^2}$, can be regarded as guiding waves for the photons; the waves themselves have no energy—there are only photons—but they are a construct whose intensity measures the average number of photons per unit volume.

We use the word "average" because the emission processes are statistical in nature. We don't specify exactly how many photons cross unit area in unit time, only their average number; the exact number can fluctuate in time and space, just as in kinetic theory of gases there are fluctuations about an average value for many quantities. We can say quite definitely, however, that the probability of having a photon cross unit area 3 meters from the source is exactly one-ninth the probability that a photon will cross unit area 1 meter from the source. In the formula $I = Nh\nu$, therefore, N is an average value and is a measure of the probability of finding a photon crossing unit area in unit time. If we equate the wave expression to the particle expression we have

$$I = \frac{1}{\mu_0 c}\overline{\mathcal{E}^2} = h\nu N$$

so that $\overline{\mathcal{E}^2}$ is proportional to N. Einstein's interpretation of $\overline{\mathcal{E}^2}$ as a probability measure of photon density then becomes clear. We expect that, as in kinetic theory, fluctuations about an average will become more noticeable at low intensities than at high intensities, so that the granular quantum phenomena would contradict the continuum classical view more dramatically there (as, for example, in the photoelectric effect discussed in Chapter 5).

In analogy to Einstein's view of radiation, Max Born proposed a similar uniting of the wave-particle duality for matter. Although this came after a wave theory for material particles, called wave mechanics, had been developed by Schrödinger, it was Born's interpretation that conceptualized the formal theory. Let us associate with matter waves not only a wavelength but also an amplitude. The function representing the de Broglie wave is called a *wave function*, signified by ψ. For particles moving in the x-direction with constant linear momentum, for example, the wave function can be described by simple sinusoidal functions, such as

$$\psi(x,t) = A \sin 2\pi\left(\frac{x}{\lambda} - \nu t\right).$$

This is analogous to

$$\mathcal{E}(x,t) = \mathcal{E}_{max} \sin 2\pi\left(\frac{x}{\lambda} - \nu t\right)$$

for the electric field of a sinusoidal electromagnetic wave of wavelength λ, and frequency ν, moving in the positive x-direction. Then $\overline{\psi^2}$ will play a role for matter waves analogous to that played by $\overline{\mathcal{E}^2}$ for waves of radiation. $\overline{\psi^2}$, the average of the square of the wavefunction of matter waves, is a measure of the average number of particles per unit volume; or, put another way, it is proportional to the probability of finding a particle in unit volume at a given place and time. Just as $\mathcal{E}$ is a function of space and time, so is ψ. And just as $\mathcal{E}$ satisfies a wave equation it turns out that so does ψ (Schrödinger's equation). The quantity $\mathcal{E}$ is a (radiation) wave associated with a photon and ψ is a (matter) wave associated with a mass particle; neither $\mathcal{E}$ nor ψ give the path of the motion but instead are associated waves that measure probability densities.

As Max Born (Ref. 7) says "According to this view, the whole course of events is determined by the laws of probability; to a state in space there corresponds a definite probability, which is given by the de Broglie wave associated with the state. A mechanical process is therefore accompanied by a wave process, the guiding wave, described by Schrödinger's equation, the significance of which is that it gives the probability of a definite course of the mechanical process. If, for example, the amplitude of the guiding wave is zero at a certain point in space, this means that the probability of finding the electron at this point is vanishingly small." Just as in the Einstein view of radiation we do not specify the exact location of a photon at a given time but specify instead by $\overline{\mathcal{E}^2}$ *the probability* of finding a photon at a certain space interval at a given time, so here in Born's view we do not specify the exact location of a particle at a given time but specify instead by $\overline{\psi^2}$ the *probability* of finding a particle in a certain space interval at a given time. And, just as we are accustomed to adding wave functions ($\mathcal{E}_1 + \mathcal{E}_2 = \mathcal{E}$) for two superposed electromagnetic waves whose resultant intensity is given by $\mathcal{E}^2$, so we add wave functions for two superposed matter waves ($\psi_1 + \psi_2 = \psi$) whose resultant intensity is given by ψ^2. That is, a *principle of superposition* applies to matter as well as to radiation. This is in accordance with the striking experimental fact that matter exhibits interference and diffraction properties, a fact that simply cannot be understood on the basis of ideas in classical mechanics. Because waves can be superposed either constructively (in phase) or destructively (out of phase), two waves can combine either to yield a resultant wave of large intensity or to cancel. But two classical particles of matter cannot combine in such a way as to cancel.

The student might accept the logic of this fusion of wave and particle

concepts but nevertheless ask whether a probabilistic or statistical interpretation is necessary. It was Heisenberg and Bohr who, in 1927, first showed
how essential the concept of probability is to the union of wave and particle
descriptions of matter and radiation. We investigate these matters in succeeding sections.

6.4 The Uncertainty Principle

The use of probability considerations is not strange to classical physics.
However, in classical physics the basic laws (such as Newton's laws) are
deterministic and statistical analysis is simply a practical device for treating
very complicated systems. According to Heisenberg and Bohr, however, the
probabilistic view is the fundamental one in quantum physics and deter-
minism must be discarded. Let us see how this conclusion is reached.

In classical mechanics the equations of motion of a system with given forces
can be solved to give us the position and momentum of a particle at all values
of the time. All we need to know are the precise position and momentum
of the particle at some value of the time $t = 0$ (the initial conditions) and
the future motion is determined exactly. This mechanics has been used with
great success in the macroscopic world, for example in astronomy, to predict
the subsequent motions of objects in terms of their initial motions, in agreement with our subsequent observations. Note, however, that in the process
of making observations the observer interacts with the system. An example
from contemporary astronomy is the precise measurement of the position of
the moon by bouncing radar from it. The motion of the moon is disturbed
by the measurement but due to the very large mass of the moon the disturbance can be ignored. On a somewhat smaller scale, as in a very well-designed
macroscopic experiment on earth, such disturbances are also usually small,
or at least controllable, and they can be taken into account accurately ahead
of time by suitable calculations. Hence, it was naturally assumed by classical
physicists that in the realm of microscopic systems the position and momentum of an object, such as an electron, could be determined precisely by
observations in a similar way. Heisenberg and Bohr questioned this assumption.

The situation is somewhat similar to that existing at the birth of relativity
theory. Physicists spoke of length intervals and time intervals, i.e., space and
time, without asking critically how we actually measured them. For example,
they spoke of the simultaneity of two separated events without even asking
how one would physically go about establishing simultaneity. In fact, Einstein
showed that simultaneity was not an absolute concept at all, as had been
assumed previously, but that two separated events that are simultaneous to
one inertial observer occur at different times to another inertial observer.
Simultaneity is a relative concept. Similarly then we must ask ourselves
critically how we actually measure position and momentum.

Can we determine by actual experiment at the same instant both the
position and momentum of matter or radiation? The answer given by quantum theory, is "not more accurately than is allowed by the Heisenberg

Uncertainty Principle." There are two parts to this principle, also called the Indeterminacy Principle (see Refs. 8 and 9). The first has to do with the simultaneous measurement of position and momentum. It states that experiment cannot simultaneously determine the exact component of momentum, p_x say, of a particle and its exact corresponding coordinate position, x. Instead, our precision of measurement is inherently limited by the measurement process itself such that

$$\Delta p_x \, \Delta x \geq \hbar/2 \qquad\qquad (6\text{-}5)$$

where the momentum p_x is known to within an uncertainty of Δp_x and the position x at the same time to within an uncertainty Δx. Here $\hbar$ (read h bar) is a shorthand symbol for $h/2\pi$, that is $\hbar = h/2\pi$ where h is Planck's constant. There are corresponding relations for other components of momentum, namely $\Delta p_y \, \Delta y \geq \hbar/2$ and $\Delta p_z \, \Delta z \geq \hbar/2$, and for angular momentum as well. It is important to realize that this principle has nothing to do with improvements in instrumentation leading to better simultaneous determinations of p_x and x. Rather, the principle says that even with ideal instruments we can never in principle do better than $\Delta p_x \, \Delta x \geq \hbar/2$. Note also that the *product* of uncertainties is involved, so that, for example, the more we modify an experiment to improve our measure of p_x, the more we give up ability to determine x accurately. If p_x is known exactly we know nothing at all about x (i.e., if $\Delta p_x = 0$, $\Delta x = \infty$). Hence, *the restriction is not on the accuracy to which x or p_x can be measured, but on the product $\Delta p_x \, \Delta x$ in a simultaneous measurement of both.*

The second part of the uncertainty principle has to do with the measurement of energy E and time t. For example, the time interval Δt in which a photon of energy uncertainty ΔE is emitted from an atom is governed by

$$\Delta E \, \Delta t \geq \hbar/2. \qquad\qquad (6\text{-}6)$$

In general, ΔE is the uncertainty in our knowledge of the energy E of a system and Δt the time interval characteristic of the rate of change in the system. This relation too sets an ideal limit to measurement of two corresponding quantities. Equation 6-6 is different, however, from Eq. 6-5 because the position and momentum variables can be measured *at a given time* and play symmetric roles whereas energy and time play different roles, the energy being a variable and the time being a parameter.

Heisenberg's relations can be shown to follow from the new wave (or quantum) mechanics, but that new mechanics was invented, after all, to explain the experiments we have already discussed (and others). Hence, the principle is grounded in experiment. We shall show shortly examples of the consistency of Eq. 6-5 and Eq. 6-6 with experiment. Notice first, however, that it is Planck's constant h that again distinguishes the quantum results from the classical ones. If h, or $\hbar$, in Eqs. 6-5 and 6-6 were zero there would be no basic limitation on our measurement at all, which is the classical view. And again it is the smallness of h that takes the principle out of the range of our ordinary experiences. This is analogous to the smallness of $\beta(= u/c)$ in the macroscopic situations taking relativity out of the range of ordinary experience. In principle, therefore, classical physics is of limited validity and in the microscopic domain it will lead to contradictions with experiments. For if we cannot determine x and p simultaneously then we cannot specify the initial conditions of motion exactly; therefore, we cannot precisely determine the future behavior of a system. Instead of making deterministic predictions we can only state the possible results of an observation, giving the relative probabilities of their occurrence. Indeed, since the act of observing a system disturbs it in a manner that is not completely predictable, the observation changes the previous motion of the system to a new state of motion which cannot be completely known.

Let us now illustrate the physical origin of the uncertainty principle. First, we use a thought experiment due to Bohr to derive Eq. 6-5. Let us say that we wish to measure as accurately as possible the position of a "point" particle, like an electron. For greatest precision we use a microscope to view the electron (Fig. 6-9a). But to see the electron we must illuminate it, for it is really the light quanta scattered by the electron that the observer sees. At this point, even before any calculations are made, we can see the uncertainty principle emerge. The very act of observing the electron disturbs it. The moment we illuminate the electron, it recoils because of the Compton effect, in a way that cannot be completely determined (see Question 18). But if we don't illuminate the electron, we don't see (detect) it. Hence the uncertainty principle refers to the measuring process itself, and expresses the fact that there is always an undetermined interaction between observer and observed; there is nothing we can do to avoid the interaction or to allow for it ahead of time. Let us try to reduce the disturbance to the electron as much as possible by using a very weak source of light. The very weakest we can get it is to assume that we can see the electron if only *one* scattered photon enters the lens of the microscope (see Fig. 6-9b). The magnitude of the momentum of the photon is $p = h/\lambda$. But the photon may have been scattered *anywhere* within the angular range 2θ subtended by the lens at the electron. This is why the interaction cannot be allowed for. The conservation equations in the Compton effect can be satisfied for *any* angle of scattering within this angular range. (See Question 18.) Hence, we see that the x-component of the momentum of the photon can vary from $+p \sin \theta$ to $-p \sin \theta$ and is uncertain after the scattering by an amount

$$\Delta p_x \simeq 2p \sin \theta = 2h/\lambda \sin \theta.$$

Because of conservation of momentum the electron must receive a recoil momentum in the x-direction that is equal to the x-momentum change in the photon and, therefore the x-momentum of the electron is uncertain by this same amount. Notice that to reduce Δp_x we should use light of longer wavelength, or use a microscope with a lens subtending a smaller angle θ.

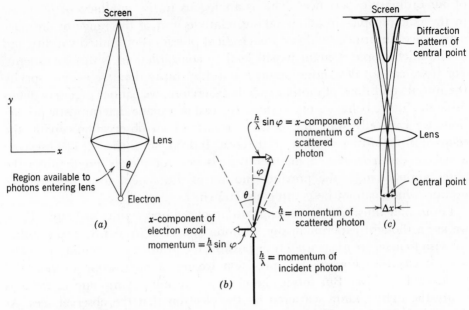

Figure 6-9. Illustrating Bohr's gamma-ray microscope thought experiment. (*a*) the geometry and coordinate axes; (*b*) the scattering of the gamma ray from the recoiling electron; (*c*) the separation Δx of points barely resolvable from the central point whose diffraction pattern image is shown.

What about the location along x of the electron? Recall that the image of a point object illuminated by light (photons) is not a point, but a diffraction pattern (see Fig. 6-9*c*). The image of the electron is "fuzzy." It is the resolving power of a microscope which determines the highest accuracy to which the electron can be located. Let us take the uncertainty Δx to be the linear separation of points in the object barely resolvable in the image. Then, from physical optics,* we have

$$\Delta x \simeq \lambda/\sin\theta.$$

The one scattered photon at our disposal must have originated then *somewhere* within this range of the axis of the microscope, so the uncertainty in the electron's location is Δx. (We can't be sure exactly where any one photon originates even though a large number of photons will show the statistical regularity of the diffraction pattern shown in the figure.) Notice that to reduce Δx we should use light of shorter wavelength, or a microscope with larger θ.

If now we take the product of the uncertainties we obtain

$$\Delta p_x \, \Delta x = \left(\frac{2h}{\lambda}\sin\theta\right)\left(\frac{\lambda}{\sin\theta}\right) = 2h > \hbar/2. \tag{6-7}$$

We cannot *simultaneously* make Δp_x and Δx as small as we wish, for the

*This result is easily derived from the result, $\sin\theta = 1.22\,\lambda/D$, of *Physics*, Sec. 44-5, Part II; see, e.g., Ref. 12.

procedure that makes one small makes the other large. For instance, if we use light of short wavelength (e.g., gamma rays) to reduce Δx (better resolution), we increase the Compton recoil and increase Δp_x, and conversely. Indeed, the wavelength λ doesn't even appear in the result. In practice an experiment might do much worse than Eq. 6-7 suggests, for that result represents the very ideal possible. We arrive at it, however, from genuinely measurable physical results, namely the Compton effect and the resolving power of a lens.

There really should be no mystery in the student's mind about our result; it is a necessary consequence of the quantization of radiation. For to have light we had to have at least one quantum of it. It is this single scattered light quantum, carrying a momentum of magnitude $p = h/\lambda$, that gives rise to an interaction between the microscope and the electron. The electron is disturbed in a manner that cannot be exactly predicted or controlled. Consequently we cannot know exactly what the coordinates and momentum of the electron will be after the interaction. If classical physics were valid, and since radiation is regarded there as continuous rather than granular, we could reduce the illumination to arbitrarily small levels and deliver arbitrarily small momentum while using arbitrarily small wavelengths for "perfect" resolution. In principle there would be no simultaneous lower limit to resolution or momentum recoil and there would be no uncertainty principle. But we cannot do this; the single photon is indivisible. Again we see, from $\Delta p_x \, \Delta x \geq \hbar/2$ that the value of Planck's constant determines the minimum uncontrollable disturbance that distinguishes quantum physics from classical physics.

Now let us consider Eq. 6-6 relating energy and time. For the case of a free particle we can obtain Eq. 6-6 from Eq. 6-5, which relates position and momentum, as follows. Consider again an electron moving along the x-axis whose energy we can write as $E = p_x^2/2m$. If p_x is uncertain by Δp_x, then the uncertainty in E is given by differentiation as $\Delta E = (2p_x/2m) \, \Delta p_x = v_x \, \Delta p_x$. Here v_x can be interpreted as the recoil velocity along x of the electron which is illuminated with light. If the time interval required for the observation of the electron is Δt, then the uncertainty in its x-position is $\Delta x = v_x \, \Delta t$. Combining $\Delta t = \Delta x/v_x$ and $\Delta E = v_x \, \Delta p_x$ we obtain $\Delta E \, \Delta t = \Delta p_x \, \Delta x$. But $\Delta p_x \, \Delta x \geq \hbar/2$. Hence, we obtain the result

$$\Delta E \, \Delta t \geq \hbar/2.$$

▶ **Example 4.** The speed of a bullet ($m = 50$ gm) and the speed of an electron ($m = 9.1 \times 10^{-28}$ gm) are measured to be the same, namely 300 m/sec, with an uncertainty of 0.01%. With what fundamental accuracy could we have located the position of each, if the position is measured simultaneously with the speed in the same experiment.

For the electron

$$p = mv = (9.1 \times 10^{-31} \text{ kg})(300 \text{ m/sec}) = 2.27 \times 10^{-28} \text{ kg-m/sec and}$$
$$\Delta p = m \, \Delta v = 0.01\% \text{ of } p = (0.0001)(2.7 \times 10^{-28} \text{ kg-m/sec})$$
$$= 2.7 \times 10^{-32} \text{ kg-m/sec so that}$$

$$\Delta x \geq \frac{h}{4\pi \, \Delta p} = \frac{6.6 \times 10^{-34} \text{ joule-sec}}{4\pi \times 2.7 \times 10^{-32} \text{ kg-m/sec}} = 2 \times 10^{-3} \text{ m} = 0.2 \text{ cm.}$$

For the bullet

$$p = mv = (0.05 \text{ kg})(300 \text{ m/sec}) = 15 \text{ kg-m/sec}$$

and

$$\Delta p = (0.0001)(15 \text{ kg-m/sec}) = 1.5 \times 10^{-3} \text{ kg-m/sec}$$

so that

$$\Delta x \geq \frac{h}{4\pi \, \Delta p} = \frac{6.6 \times 10^{-34} \text{ joule-sec}}{4\pi \times 1.5 \times 10^{-3} \text{ kg-m/sec}} = 3 \times 10^{-32} \text{ m}.$$

Hence, for macroscopic objects such as bullets the uncertainty principle sets no practical limit to our measuring procedure, Δx in this example being about 10^{-17} times the diameter of a nucleus. But, for microscopic objects such as electrons, there are practical limits, Δx in this example being about 10^7 times the diameter of an atom.

Example 5. You wish to localize the y-coordinate of an electron by having it pass through a narrow slit. By considering the diffraction of the associated wave, show that as a result you introduce an uncertainty in the momentum of the electron such that $\Delta y \, \Delta p_y \geq \hbar/2$.

In Fig. 6-10 we show the diffraction pattern formed on a screen by a parallel beam of monoenergetic electrons that first passed through a single slit placed in the path of the beam. From the wave point of view we can regard this as the passage of a plane monochromatic wave of wavelength λ through a single slit of width a. From physical optics (*Physics*, Part II, Sec. 44-2) we know that the angle θ to the first diffraction minimum is given by $a \sin \theta = \lambda$. From the particle point of view, the diffraction pattern gives the statistical distribution on the screen of a large number of electrons of incident momentum p that may be deflected up or down on passing through the slit.

Hence, the wave particle duality manifests itself here. On counting the arrival of individual electrons on the screen, the particle aspect is revealed. On looking at the total distribution on the screen, i.e., the diffraction pattern, the wave aspect is revealed.

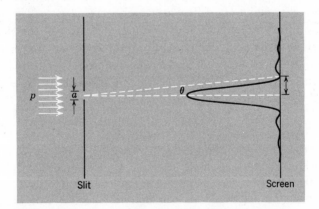

Figure 6-10. Example 5. Electrons in a parallel beam, each electron of momentum p, are diffracted at a single slit of width a, forming a diffraction pattern on the screen. Individual electrons can be detected by counters, for example. Although there is uncertainty in predicting the exact position and momentum of an individual electron diffracted at the slit, there is a predictable statistical regularity in the pattern formed by a large number of electrons.

Each electron is regarded as being diffracted independently. The probability that an electron hits some point on the screen is determined by ψ^2, the square of the amplitude of the guiding wave. Where ψ^2 is zero, no electrons are observed; where ψ^2 is large, many electrons are observed. At high incident intensities the predicted diffraction pattern is easily observed. At low intensities, a short exposure will give an average pattern like the high intensity one but with statistical fluctuations, whereas a long exposure will clearly reveal the high intensity pattern.

For a *single* electron arriving at the screen, therefore, we do not know exactly where it passes through the slit, only that it did pass through. Hence, the uncertainty in its y-coordinate at the slit is of the order of the slit width, i.e.,

$$\Delta y \simeq a.$$

As for the y-component of the momentum of a single electron at the slit, again we are uncertain of its value but we know from the existence of the diffraction pattern that electrons do acquire vertical momentum on being deflected there. If we consider θ to be an *average* deflection angle (we don't know exactly where a single electron will hit the screen), then the uncertainty Δp_y in the y-component of momentum of a single electron is of the order of $p \sin \theta$, i.e.,

$$\Delta p_y \simeq p \sin \theta.$$

Hence,

$$\Delta y \, \Delta p_y \simeq ap \sin \theta.$$

But $p = h/\lambda$ and $a \sin \theta = \lambda$, so that

$$\Delta y \, \Delta p_y \simeq \frac{\lambda h}{\lambda} \simeq h$$

and

$$\Delta y \, \Delta p_y \geq \frac{\hbar}{2}. \blacktriangleleft$$

6.5 A Derivation of the Uncertainty Principle

We can derive the uncertainty relations in a somewhat formal way by combining the de Broglie Einstein relations, $p = h/\lambda$ and $E = h\nu$, with certain properties that are universal to all waves. We do this in succeeding paragraphs. But first let us get a feeling for the derivation by considering wave phenomena.

Let λ be the wavelength of a de Broglie wave associated with a particle. We can picture a definite (monochromatic) wavelength in terms of a sinusoidal wave extending in one dimension over all values of x, that is, an infinitely long wave train. The wave function can be described as $\psi = A \sin 2\pi(x/\lambda - \nu t)$ or $\psi = A \cos 2\pi(x/\lambda - \nu t)$. Now, if λ is definite there is no uncertainty $\Delta \lambda$ and the associated particle momentum $p = h/\lambda$ is also definite; therefore, $\Delta p_x = 0$. But in such a (monochromatic) wave the amplitude of the de Broglie wave does not vary; it is the same over the entire range of x. Therefore, the probability of finding the associated particle is not concentrated in a particular range of x. In other words, the location of the particle is completely unknown; it can be anywhere, so that $\Delta x = \infty$. Analogous statements are that since E equals $h\nu$ and since the frequency is definite, then $\Delta E = 0$. But to be sure that the amplitude of the wave is constant in time we must observe the wave for an infinite time, so that $\Delta t = \infty$.

In order to have a wave whose amplitude varies with x or t, we must superpose several sinusoidal waves of different wavelengths or frequencies. In the simple case of two such waves we obtain the familiar phenomena of beats (see Part I, Sec. 20-6), the amplitude varying in a regular way throughout space or time. If we wish to construct a wave train having a finite extent in space (i.e., a definite beginning and end) so as to obtain a pulse, such as a photon, then we must synthesize waves having a continuous spectrum of wave-lengths within a range $\Delta\lambda$. The amplitude of such a pulse will be zero everywhere outside a region of extent Δx.

There is a well-known procedure, using Fourier theory (see, e.g., Ref. 10), in which an infinitely large number of monochromatic waves, differing infinitesimally in λ and ν, are combined to give a wave group, such as a single pulse, which has a finite extent in space. The compo-nent waves, because of their phase differences and chosen amplitudes, interfere destructively to give a zero resultant everywhere outside of a limited region, whereas within this region they combine con-structively to form the pulse. For a pulse, or wave train, of particular extension Δx in space one must choose a particular spread of wave-lengths $\Delta\lambda$ in the component waves (see e.g. Fig. 6-11). The relation, given in terms of reciprocal wavelength $\kappa = 1/\lambda$, is $\Delta x\,\Delta\kappa \geq 1/4\pi$ or

$$\Delta x\, \Delta\left(\frac{1}{\lambda}\right) \geq \frac{1}{4\pi}. \tag{6-8}$$

If, for example, a light pulse of spatial extent Δx is sent through an optical system that is dispersive, the system will behave exactly as if it had received a superposition of sinusoidal waves of wavelength range given by $\Delta(1/\lambda)$ of Eq. 6-8 and Fourier analysis.

A group of waves that has a finite extension in space also has a finite duration at a point of observation. If Δt is the duration of the pulse, then its time of arrival can be determined to within a limited precision Δt. To construct a group of waves having a finite duration in time, we must synthesize waves having a continuous spectrum of frequencies within a range $\Delta\nu$. The relation between $\Delta\nu$ and Δt is the same as that between $\Delta\kappa$ and Δx (see, e.g., Ref. 10). That is,

$$\Delta\nu\, \Delta t \geq \frac{1}{4\pi}. \tag{6-9}$$

An electromagnetic pulse of duration Δt has no periodicity, but if such a pulse is sent through a circuit whose behavior is frequency dependent the circuit will respond precisely the same as it would if it had received a superposition of sinusoidal waves of frequency range $\Delta\nu$ given by Eq. 7-9 and Fourier analysis.

Equations 6-8 and 6-9 are universal properties of all waves. From these equations we see that the smaller the spread in reciprocal wave-

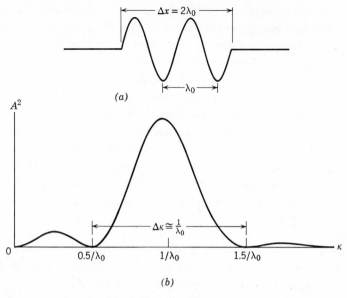

(a)

(b)

Figure 6-11. (a) A wave pulse of length Δx, consisting of two cycles of a sine wave of wavelength λ_0. (b) The pulse can be constructed by superimposing many sine waves of different amplitudes A and wavelengths λ. Note that the major contribution comes from waves with wavelengths near λ_0, but waves with wavelengths substantially different from λ_0 must also be present to insure that the pulse has zero amplitude everywhere outside the region Δx. It is customary to plot such amplitude distributions as squared-amplitudes A^2 against reciprocal wavelength $\kappa = 1/\lambda$, as has been done here. All the waves which make a significant contribution to the superposition have reciprocal wavelengths lying in a range $\Delta\kappa \cong 1/\lambda_0$, so $\Delta x\,\Delta\kappa \cong 2$, confirming the relation $\Delta x\,\Delta\kappa \geq 1/4\pi$.

lengths or frequencies used the longer the extent of the wave pulse and its duration at a point. We have already seen that a wave of single wavelength and frequency has infinite extent in space ($\Delta x = \infty$) at a given time and infinite duration in time ($\Delta t = \infty$) at a given point. To construct a sharp pulse, one that is confined to an infinitesimally small region of space ($\Delta x = 0$), we must superpose sinusoidal waves whose wavelengths include *all* values from zero to infinity ($\Delta\kappa = \infty$). We know nothing at all about the frequency of such a pulse ($\Delta\nu = \infty$) but its time of arrival is known exactly ($\Delta t = 0$).

To obtain the Heisenberg uncertainty relations, we now combine the de Broglie relation $p = h/\lambda$ with Eq. 6-8 to obtain $\Delta x\,\Delta(1/\lambda) = \Delta x\,\Delta(p/h) \geq 1/4\pi$ or

$$\Delta x\,\Delta p \geq \frac{\hbar}{2} \qquad (6\text{-}5)$$

and combine the Einstein relation $E = h\nu$ with Eq. 6-9 to obtain $\Delta\nu\,\Delta t = \Delta(E/h)\,\Delta t \geq 1/4\pi$ or

$$\Delta E\,\Delta t \geq \frac{\hbar}{2}. \qquad (6\text{-}6)$$

As an illustration of Eq. 6-6 consider the excitation of an atom by radiation. Suppose we excite an atom by having it absorb incident monochromatic light (no uncertainty in ν or E) and we ask what is the time of the absorption. From $\Delta E \; \Delta t \geq h/4\pi$ the answer is $\Delta t = \infty$, so that the time or absorption is completely indeterminate. Experimentally, to determine the moment of absorption, we would have to mark the original wave train somehow to follow it; in practice we might use a shutter mechanism to interrupt it and "chop" the light. But this act of measurement disturbs the assumed monochromatic nature of the light; it is now a finite pulse which necessarily has a frequency spread $\Delta \nu$ and a corresponding uncertainty ΔE in the energy. The narrower the pulse and the sharper Δt, the greater the spread in energy ΔE.

▶ **Example 6.** An atom can radiate at any time after it is excited. It is found that in a typical case the average excited atom has a lifetime of about 10^{-8} secs. That is, during this period it emits a photon and is deexcited.

(*a*) What is the minimum uncertainty $\Delta \nu$ in the frequency of the photon? From Eq. 6-9 we have

$$\Delta \nu \; \Delta t \geq \frac{1}{4\pi}$$

or

$$\Delta \nu \geq \frac{1}{4\pi \, \Delta t}.$$

With $\Delta t = 10^{-8}$ secs we obtain $\Delta \nu \geq 8 \times 10^{6}$ sec^{-1}.

(*b*) Such photons from sodium atoms appear in a spectral line centered at $\lambda = 5890$ Å. What is the fractional width of the line, $\Delta \nu / \nu$? For $\lambda = 5890$ Å, we obtain

$$\nu = \frac{c}{\lambda} = \frac{3 \times 10^{10} \text{ cm/sec}}{5890 \times 10^{-8} \text{ cm}} = 5.1 \times 10^{14} \text{ sec}^{-1}.$$

Hence

$$\frac{\Delta \nu}{\nu} = \frac{8 \times 10^{6} \text{ sec}^{-1}}{5.1 \times 10^{14} \text{ sec}} = 1.6 \times 10^{-8}$$

or about two parts in 100 million.

This is the so-called *natural width* of the spectral line. The line is much broader in practice because of the Doppler broadening and pressure broadening due to the motions and collisions of atoms in the source.

(*c*) Calculate the uncertainty ΔE in the energy of the excited state of the atom.

The energy of the excited state is not precisely measurable because only a finite time is available to make the measurement. That is, the atom does not stay in an excited state for an indefinite time but decays to its lowest energy state, emitting a photon in the process. The uncertainty in energy of the photon equals the uncertainty in energy of the excited state of the atom, in accordance with the energy conservation principle. From Eq. 6-6, with Δt equal to the mean lifetime of the excited state, we have

$$\Delta E \geq \frac{h/4\pi}{\Delta t} = \frac{h}{4\pi \, \Delta t} = \frac{6.63 \times 10^{-34} \text{ joule-sec}}{4 \times 10^{-8} \text{ sec}}$$

$$= \frac{4.14 \times 10^{-15} \text{ ev-sec}}{4 \times 10^{-8} \text{ sec}} \; \frac{1}{3} \times 10^{-7} \text{ ev}.$$

This agrees, of course, with the value obtained from part (a) by multiplying the uncertainty in photon frequency $\Delta \nu$ by h to obtain $\Delta E = h \, \Delta \nu$.

The uncertainty in energy of an excited state is usually called the *energy level width* of the state.

(d) From the previous results determine, to within an accuracy ΔE, the energy E of the excited state of a sodium atom, relative to its lowest energy state, that emits a photon whose wavelength is centered at 5890 Å.

We have

$$\frac{\Delta \nu}{\nu} = \frac{h \, \Delta \nu}{h \nu} = \frac{\Delta E}{E}.$$

Hence,

$$E = \frac{\Delta E}{(\Delta \nu / \nu)} = \frac{\frac{1}{3} \times 10^{-7} \text{ eV}}{1.6 \times 10^{-8}} = 2.1 \text{ ev},$$

in which we have used the results of the calculations in parts (b) and (c). ◀

6.6 Interpretation of the Uncertainty Principle

If we take the uncertainty principle as fundamental, then we can understand why it is possible for both light and matter to have a dual, wave-particle nature. Max Born (Ref. 7) has stated the connection as follows: "It is just the limited feasibility of measurements that define the boundaries between our concepts of a particle and a wave. The corpuscular (particlelike) description means at bottom that we carry out the measurements with the object of getting exact information about momentum and energy relations (e.g., in the Compton effect), while experiments which amount to determination of place and time we can always picture to ourselves in terms of the wave representation (e.g., passage of electrons through thin foils and observations of deflected beam) . . . (But) just as every determination of position carries with it an uncertainty in the momentum, and every determination of time an uncertainty in the energy, so the converse is also true. The more accurately we determine momentum and energy, the more latitude we introduce into the position of the particle and time of an event."

If we try experimentally to determine whether radiation is a wave or a particle, for example, we find that an experiment that forces radiation to reveal its wave character strongly suppresses its particle character. If we modify the experiment to bring out the particle character, its wave character is suppressed. We can never bring the wave and the particle view face to face in the same experimental situation. Matter and radiation are like coins that can be made to display either face at will but not both simultaneously. This, of course, is the essence of Bohr's principle of complementarity, the ideas of wave and of particle complementing rather than contradicting one another.

We have also seen that the probability idea emerges directly from the uncertainty principle. In classical mechanics, if at any instant we know exactly the position and momentum of each particle in an isolated system, then we can predict the exact behavior of the particles of the system for all future time. In quantum (or wave) mechanics, however, the uncertainty principle shows us that it is impossible to do this for systems involving small distances and momenta because it is impossible to know, with the required accuracy, the instantaneous positions and momenta of the particles of such a system.

As a result, we are able to make predictions only of the *probable* behavior of these particles.

Quantum theory therefore, abandons strict causality in individual events in favor of a fundamentally statistical interpretation of their collective regularity. The commonly accepted statistical view (the Copenhagen interpretation) is given by Max Born (Ref. 7), whom we can paraphrase as follows. The law of causation, according to which the course of events in an isolated system is completely determined by the state of the system at $t = 0$, loses its validity in quantum theory. If we view processes from a wave-particle picture, then because this duality essentially involves indeterminacy we are forced to abandon a deterministic theory. Likewise, if we examine processes analytically in quantum theory, we must describe the instantaneous state of a system by a wave function ψ. It is true that ψ satisfies a differential equation and therefore changes with time from its form at $t = 0$ in a causal way. But physical significance is confined to ψ^2, the square of the amplitude, and this only partially defines ψ. That is, the initial value of the wavefunction ψ is necessarily not completely definable. If we cannot in principle know the initial state exactly it is empty to assert that events develop in a strictly causal way, for physics is then in the nature of the case indeterminate and statistical.

Heisenberg stated the matter succinctly as follows:

"We have not assumed that the quantum theory, as opposed to the classical theory, is essentially a statistical theory, in the sense that only statistical conclusions can be drawn from exact data. . . . In the formulation of the causal law, namely, 'If we know the present exactly, we can predict the future,' it is not the conclusion, but rather the premise which is false. We *cannot* know, as a matter of principle, the present in all its details."

The connection between probability and the wave function becomes apparent in single or double slit interference experiments. Consider a parallel beam of electrons incident on a double slit. The motion of the electrons is governed by the de Broglie waves associated with them. The original wavefront is split into two coherent wavefronts by the slits and these overlapping wavefronts produce the interference fringes on the screen. The intensity along the screen follows the usual interference pattern and corresponds to ψ^2, where ψ is the wave function. But the same intensity variation along the screen can be obtained by counting the numbers of electrons arriving at different parts of the screen in a given time. The probability that a single electron will arrive at a given place is just the ratio of the number that arrive there to the total number. In other words, the statistical distribution of a large number of electrons yields the interference pattern predicted by the wave function. The intensity ψ^2 of the de Broglie wave at a particular position is then a measure of the probability that the electron is located at that position.

6.7 Conclusion

To summarize, we have seen that physical measurement necessarily involves interaction between the observer and the system being observed. Matter and radiation are the entities available to us for such measurements. The relations $p = h/\lambda$ and $E = h\nu$ apply to matter and to radiation, being the expression of the wave-particle duality. When we combine these relations with the properties universal to all waves we obtain the uncertainty relations. Hence, the uncertainty principle is a necessary consequence of this duality, that is, of the de Broglie-Einstein relations. And the uncertainty principle itself is the basis for the Heisenberg-Bohr contention that the probabilistic view is basic to quantum physics.

QUESTIONS

1. Why is the wave nature of matter not more apparent to us in our daily observations?

2. Does the de Broglie wavelength apply only to "elementary particles" such as an electron or neutron, or does it apply as well to compound systems of matter (such as a molecule) having internal structure? Give examples.

3. If, in the de Broglie formula, we let $m \to \infty$, do we get the classical result for particles of matter?

4. Can the de Broglie wavelength of a particle be smaller than a linear dimension of the particle? . . . larger? Is there any relation necessarily between such quantities?

5. How can electron diffraction be used to study properties of the surface of a solid?

6. How does one account for regularly reflected beams in diffraction experiments with electrons and atoms?

7. Does the Bragg formula have to be modified for electrons to account for the refraction of electron waves at the crystal surface?

8. Explain qualitatively how a magnetic lens system for electrons in an electron microscope works.

9. Why is an X-ray microscope difficult to construct?

10. Do electron diffraction experiments give different information about crystals than can be obtained from X-ray diffraction experiments? . . . from neutron diffraction experiments? Discuss.

11. Could crystallographic studies be carried out with protons or heavier ions? Explain.

12. Discuss the analogy: physical optics $\to$ geometrical optics
and: wave mechanics $\to$ classical mechanics.

13. Why does a Laue pattern display "dots" whereas a Debye-Scherrer pattern displays "circles"?

14. Is an electron a particle? Is it a wave? Explain.

15. Considering electrons and photons to be particles, how are they basically different from each other?

16. Does the de Broglie wavelength associated with a particle depend on the inertial reference frame of the observer? What effect does this have on the wave-particle duality? (See Ref. 11.)

17. (*a*) Give examples of how the process of measurement disturbs the system being measured. (*b*) Can the disturbances be taken into account ahead of time by suitable calculations?

18. Show the relation between the uncontrollable nature of the Compton recoil in Bohr's gamma-ray microscope experiment and the fact that there are four unknowns and only three conservation equations in the Compton effect.

19. Argue from the Heisenberg uncertainty principle that the lowest energy of an oscillator cannot be zero. (See Prob. 28.)

20. Discuss similarities and differences between a matter wave an an electromagnetic wave.

21. Is there a contradiction between the statement that the predictions of wave mechanics are exact and the fact that the information derived is of a statistical character? Explain.

22. Games of chance contain events which are ruled by statistics. Do such games violate the strict determination of individual events. Do they violate cause and effect?

PROBLEMS

1. A bullet of mass 40 gm travels at 1000 m/sec. (*a*) What wavelength can we associate with it? (*b*) Why does the wave nature of the bullet not reveal itself through diffraction effects?

2. The wavelength of the yellow spectral emission of sodium is 5896 Å. At what kinetic energy would an electron have the same de Broglie wavelength?

3. An electron and a photon each have a wavelength of 2.0 Å. What are their (*a*) momenta (in Mev/*c*) and (*b*) total energies (in Mev)? (*c*) Compare the *kinetic* energy of the electron to the energy of the photon.

4. Thermal neutrons have an average kinetic energy $\frac{3}{2}kT$ where T is room temperature, 300°K. Such neutrons are in thermal equilibrium with normal surroundings. (*a*) What is the average energy in ev of a thermal neutron? (*b*) What is the corresponding de Broglie wavelength?

5. Compare the wavelength of a 1 Mev (kinetic energy) electron, a 1 Mev photon, and a 1 Mev (kinetic energy) neutron.

★6. (*a*) Show that the de Broglie wavelength of a particle, of charge e and rest mass m_0, moving at relativistic speeds is given as a function of the accelerating potential V as

$$\lambda = \frac{h}{\sqrt{2m_0eV}}\left(1 + \frac{eV}{2m_0c^2}\right)^{-1/2}.$$

(*b*) For an electron, what accelerating voltages will keep the classical result $\lambda = h/\sqrt{2m_0eV}$, accurate to within 1%?

★7. Determine at what energy, in electron volts, the non-relativistic expression for the de Broglie wavelength will be in error by one percent for (*a*) an electron and (*b*) a neutron. (See Prob. 6.)

8. The 50 Gev (= 50 × 10⁹ ev) electron accelerator at Stanford provides an electron beam of very short wavelength, suitable for probing the fine details of nuclear structure by scattering experiments. What is this wavelength and how does it compare to the size of an average nucleus (radius ≃ 1.5 × 10⁻¹⁵ m)? (*Hint*. At these energies it is simpler to use the extreme relativistic relationship between momentum and energy, namely $p = E/c$. This is the same relationship used for light and is justified whenever the kinetic energy of a particle is very much greater than its rest energy m_0c^2, as in this case.)

9. Make a plot of de Broglie wavelength against kinetic energy for (*a*) electrons and (*b*) protons. Restrict the range of energy values to those in which classical mechanics applies reasonably well. A convenient criterion is that the maximum kinetic energy on each plot be only about, say 5% of the rest energy m_0c^2 for the particular particle.

10. In the experiment of Davisson and Germer (*a*) show that the second- and third-order diffracted beams corresponding to the strong maximum of Fig. 6-2 cannot occur and (*b*) find the angle at which the first-order diffracted beam would occur if the accelerating potential were changed from 54 to 60 volts? (*c*) What accelerating potential is needed to produce a second-order diffracted beam?

11. The principal planar spacing in a potassium chloride crystal is 3.14 Å. Compare the angle for first order Bragg reflection from these planes of electrons of kinetic energy 40 kev to that of 40 kev photons. (See Prob. 6 and criterion of Prob. 9 for possible relativistic considerations.)

12. In Example 3, use the relativistic expression for momentum and show that $\lambda = 0.037$ Å.

13. Electrons incident on a crystal suffer refraction due to an attractive potential of about 15 volts that crystals present to electrons (due to the ions in the crystal lattice). If the angle of incidence of an electron beam is 45° and the electrons have an incident energy of 100 ev (*a*) what is the angle of refraction and (*b*) what is the wavelength of the electron in the crystal?

14. What accelerating voltage would be required for electrons in an electron microscope to obtain the same ultimate resolving power as that which could be obtained from a "gamma-ray microscope" using 0.2 Mev gamma rays?

15. The highest achievable resolving power of a microscope is limited only by the wavelength used; that is, the smallest detail that can be separated is about equal to the wavelength. Suppose one wishes to "see" inside an atom. Assuming the atom to have a diameter of 1.0 Å, this means that we wish to resolve detail of separation about 0.1 Å. (*a*) if an electron microscope is used, what minimum kinetic energy of electrons is needed? (*b*) If a light microscope is used, what minimum energy of photons is needed? In what region of the electromagnetic spectrum are these photons? (*c*) Which microscope seems more practical for this purpose? Explain.

16. Show that for a free particle the uncertainty relation can also be written as

$$\Delta\lambda \, \Delta x \geq \frac{\lambda^2}{4\pi}$$

where Δx is the uncertainty in location of the wave and $\Delta\lambda$ the simultaneous uncertainty in wavelength.

17. If $\Delta\lambda/\lambda = 10^{-7}$ for a photon, what is the simultaneous value of Δx for (*a*) a gamma ray, $\lambda = 5.00 \times 10^{-4}$ Å? (*b*) An X ray, $\lambda = 5.00$ Å? (*c*) a light ray, $\lambda = 5000$ Å? (See Prob. 16.)

18. In a repetition of Thomson's experiment for measuring e/m for the electron, a beam of 10^4 ev electrons is collimated by passage through a slit of width 0.50 mm. Why is the beamlike character of the emergent electrons not destroyed by diffraction of the electron wave at this slit?

★**19.** A 1 Mev electron leaves a track in a cloud chamber. The track is a series of water droplets each about 10^{-5} meter in diameter. Show, from the ratio of the uncertainty in transverse momentum to the momentum of the electron, that the electron path should not noticeably differ from a straight line.

20. Show that if the uncertainty in the location of a particle is about equal to its de Broglie wavelength then the uncertainty in its velocity is about equal to its velocity.

21. The statement is often made that the momentum p is at least as large as the uncertainty in momentum Δp; or, in other words, the maximum value of Δp cannot exceed p. This is correct if for p one uses the root mean square value of the momentum. Is this plausible? Explain.

22. (a) Show that the smallest possible uncertainty in the position of an electron whose speed is given by $\beta = v/c$ is

$$(\Delta x)_{\min} = \frac{h}{4\pi m_0 c}(1 - \beta^2)^{1/2} = \frac{\lambda_c}{4\pi}\sqrt{1 - \beta^2}$$

where λ_c is the Compton wavelength $h/m_0 c$. (*Hint.* See Prob. 21.) (b) What is the meaning of $\beta = 0?\ldots$ of $\beta = 1$?

23. A microscope using photons is employed to locate an electron in an atom to within a distance of 0.2 Å. What is the uncertainty in the velocity of the electron located in this way? (*Hint.* The velocity is nonrelativistic).

24. (a) Consider an electron whose position is somewhere in an atom of diameter 1 Å. What is the uncertainty in the electron's momentum? Is this consistent with the binding energy of electrons in atoms as measured, for example, by ionization potentials? (b) Imagine an electron to be somewhere in a nucleus of diameter 10^{-12} cm. What is the uncertainty in the electron's momentum? Is this consistent with the binding energy of nuclear constituents? (c) Consider now a neutron, or a proton, to be in such a nucleus. What is the uncertainty in the neutron's (or proton's) momentum? Is this consistent with the binding energy of nuclear constituents?

25. The lifetime of an excited state of a nucleus is usually about 10^{-12} sec. What is the uncertainty in energy of the gamma ray photon emitted?

26. What is the approximate lifetime of an atomic state whose emission wavelength $\lambda = 5000$ Å is known to a precision of $\Delta\lambda/\lambda = 10^{-7}$.

27. A radar pulse of duration 0.1×10^{-6} secs is used to determine the distance from source to distant object by detection of the returning reflected pulse. Use a signal velocity c and show that the calculation of distance is uncertain by about 15 meters.

★28. The energy of a linear harmonic oscillator is $E = p_x^2/2m + kx^2/2$. (a) Show, using the uncertainty relation, that this can be written as

$$E = \frac{h^2}{32\pi^2 m x^2} + \frac{kx^2}{2}.$$

(b) Then show that the minimum energy of the oscillatior is $h\nu/2$ where

$$\nu = \frac{1}{2\pi}\sqrt{\frac{k}{m}}$$

is the oscillator frequency.
(*Hint.* Classically the minimum energy is zero. This requires that x and p_x simultaneously be zero, but our knowledge of them is limited by $\Delta x\,\Delta p_x = h/4\pi$. Therefore we must get E in terms of Δx (or Δp_x), as in part (a), and minimize E with respect to Δx (or Δp_x) in part (b).)

REFERENCES

1. Louis de Broglie, "Investigations on Quantum Theory" (the text of his doctoral thesis) in *Selected Readings in Physics—Wave Mechanics,* by Gunther Ludwig, Pergamon Press, 1968.

2. Karl K. Darrow, "Davisson and Germer," *Scientific American* (May 1948).

3. An historical account of Thomson's experiments is given by Sir George Thomson, "Early Work in Electron Diffraction," *Am. J. Phys.*, 821 (1961), and "The Early History of Electron Diffraction," *Contemporary Physics, 9,* 11 (January 1968).

4. Max Jammer, *The Conceptual Development of Quantum Mechanics,* McGraw-Hill Book Company, 1966.

5. L. Marton, "Electron Interferometer," *Phys. Rev.* **85,** 6, 1057 (1952).

6. Albert V. Crewe, "A High-Resolution Scanning Electron Microscope," *Scientific American,* April 1971, p. 26. S. Kimoto and J. C. Russ, "The Characteristics and Applications of the Scanning Electron Microscope," *American Scientist,* **57,** 112–113 (1969); W. C. Nixon, "Scanning Electron Microscopy," *Contemporary Physics,* **10,** 1 71–96 (1969); V. E. Cosslett, "The High Voltage Electron Microscope," *Contemporary Physics,* **9,** 4, 333–354 (1968).

7. Max Born, *Atomic Physics,* Hafner Pub. Co., 7th ed., 1962.

8. R. Furth, "The Limits of Measurement," *Scientific American* (July 1950).

9. George Gamow, "The Principle of Uncertainty," *Scientific American* (January 1958).

10. R. Bracewell, *The Fourier Transform and Its Applications,* McGraw-Hill, New York, 1965.

11. P. C. Peters, "Consistency of the de Broglie Relations with Special Relativity," *Am. J. Phys.*, **38,** 7, 931 (1970).

12. Born and Wolf, *Principles of Optics,* Pergamon Press, New York, 1959.

CHAPTER 7

Early Quantum Theory of the Atom

7.1 Models of the Atom

After the discovery of the electron by J. J. Thomson in 1897, many experiments were designed to determine the internal structure of atoms. The charge e on an electron was known and the electronic mass was found to be several thousand times smaller than the mass of the lightest atoms. Experiments, such as the photoelectric effect and the scattering of X rays by atoms, indicated that an atom does contain electrons and that the number Z of electrons per atom was about equal to half the atomic weight number A of the atom. On the assumption that the entire negative charge of an atom resides in its electrons, it was concluded that a neutral atom contains a positive charge whose magnitude Ze equals that of the electrons and that most of the mass of the atom must be associated with this positive charge. The question then arose as to how the electrons and the positive charge are arranged in an atom.

From the density of solid matter it had been estimated that atoms had an extent of about 10^{-10} m and from other considerations that electrons had an extent of about 10^{-15} m (see Prob. 1). J. J. Thomson constructed a model of the atom as a sphere of positive charge with electrons, regarded as practically point charges, embedded in it. This "plum pudding" model was elaborated upon by Thomson in an attempt to account for the periodicity of the chemical properties of the elements. He tried to construct stable arrangements of the electron "plums" in the positively charged "pudding" (see Fig. 7-1) and assumed that the emission of radiation by atoms was due to induced oscillations of the electrons about their equilibrium positions.

▶ **Example 1.** (*a*) In Thomson's model of the hydrogen atom, he assumed that there is one electron inside a spherical region of uniform positive charge density ρ. Show that its motion will be simple harmonic.

Let the electron be initially at rest at a distance a from the center ($a < r$). From

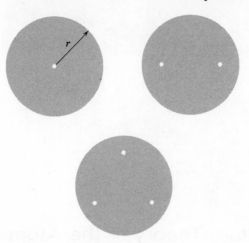

Figure 7-1. Thomson's model for atoms containing one, two, and three electrons, respectively. The electron "plums" are embedded in a positively charged spherical "pudding" of radius r.

Gauss' law, we know that we can calculate the force on it by using the formula for Coulomb's law. We obtain

$$F_a = -\frac{1}{4\pi\varepsilon_0}\left(\frac{4}{3}\pi a^3\rho\right)\frac{e}{a^2} = -\left(\frac{\rho e}{3\varepsilon_0}\right)a,$$

where $(4/3)\pi a^3\rho$ is the net positive charge in a sphere of radius a. Hence, we can write $F_a = -ka$ where the constant $k = \rho e/3\varepsilon_0$. This is the condition on a force responsible for simple harmonic motion, namely that it is proportional to the displacement but oppositely directed (see Part I, Sec. 15-2).

(*b*) Let the total positive charge equal that of the electron and be distributed in a spherical region of radius $r = 1.0 \times 10^{-10}$ m. Find the force constant k and the frequency of the motion of the electron.

We have

$$\rho = \frac{e}{(4/3)\pi r^3}$$

so that

$$k = \frac{\rho e}{3\varepsilon_0} = \frac{e}{(4/3)\pi r^3}\frac{e}{3\varepsilon_0} = \frac{e^2}{4\pi\varepsilon_0 r^3} = 2.3 \times 10^2 \text{ nt/m}.$$

The frequency of the simple harmonic motion is given by

$$\nu = \frac{1}{2\pi}\sqrt{\frac{k}{m}} = \frac{1}{2\pi}\sqrt{\frac{2.3 \times 10^2 \text{ nt/m}}{9.11 \times 10^{-31} \text{ kg}}} = 2.5 \times 10^{15} \text{ sec}^{-1}.$$

Assuming (in analogy to radiation emitted by electrons oscillating in an antenna) that the radiation emitted by the atom has this same frequency, it will correspond to a wavelength

$$\lambda = \frac{c}{\nu} = \frac{3.0 \times 10^8 \text{ m/sec}}{2.5 \times 10^{15}/\text{sec}} = 1.2 \times 10^{-7} \text{ m} = 1200 \text{ Å},$$

in the far ultraviolet portion of the electromagnetic spectrum. It is easy to show (see Prob. 2) that an electron moving in a stable circular orbit of any radius inside this Thomson atom has the same frequency.

Of course, a different assumed radius of the sphere of positive charge would give a different emission frequency (see Prob. 3). But the fact that the Thomson model hydrogen atom has only one characteristic frequency conflicts with the large number of frequencies observed in the spectrum of hydrogen. ◀

Thomson's model had to be given up completely when it was found to conflict with measurements on the scattering of alpha particles by atoms. It was Ernest Rutherford (1871–1937), a former student of Thomson's, who suggested such measurements, analyzed them, and in 1911 (see Ref. 1) proposed a new model that accounted for the observations. In his "nuclear" model, Rutherford assumed that all the positive charge of the atom, and all its mass exclusive of the mass of the electrons, is concentrated in a very small region at the center of the atom called the *nucleus*. The electrons, no longer embedded in the positive charge, were pictured as orbiting about the nucleus in planetary fashion.

Rutherford had already been awarded the Nobel prize in 1908 for his "investigations in regard to the decay of elements and . . . the chemistry of radioactive substances." He was a talented, hard-working physicist with enormous drive and self-confidence. In a letter written later in life, the then Lord Rutherford wrote "I've just been reading some of my early papers and, you know, when I'd finished, I said to myself, 'Rutherford, my boy, you used to be a damned clever fellow'." Although pleased at winning a Nobel prize he was not happy that it was a prize in chemistry rather than one in physics (any research in the elements was then considered chemistry). In his speech accepting the prize he noted that he had observed many transformations in his work with radioactivity but never had seen one as rapid as his own, from physicist to chemist.*

Alpha particles were already known to Rutherford to be doubly ionized helium atoms (i.e., He atoms with two electrons removed), emitted spontaneously from several radioactive materials at high speed. Rutherford recognized that he could use these particles as a probe to examine the structure of matter. In Fig. 7-2 we show schematically an arrangement that he used to study the scattering of alpha particles on passing through thin foils of various substances. A lead block containing a radioactive source of α-particles emits a narrow beam of alphas into a vacuum chamber at the center of which is a very thin metallic target foil. The alphas pass through the foil with little loss of energy, but may be deflected from their incident path by the electrical forces exerted on them by the atoms in the foil. A single "scattered" alpha particle can be detected by observing with a microscope the flash of light it produces on a fluorescent screen placed in its path. The detecting screen and microscope can be rotated to observe different angles ϕ of scattering and, during an experiment, the observer records the rate at which light flashes

*See Ref. 2 for an interesting discussion of Rutherford and his work.

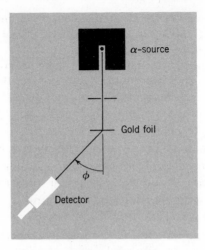

Figure 7-2. Schematic of α-particle scattering arrangement used in Rutherford's laboratories. The detector can be rotated to observe scattering at different angles ϕ.

are produced for each of many different angular positions of the detector. This reveals that an initially parallel beam of alphas incident on the foil emerges from the foil as a diverging beam. Careful measurements of the scattering should reveal information about the nature of the forces encountered by the alphas in passing through the foil, and this, in turn, might reveal the actual arrangement of the positive and negative charge in the atoms making up the foil.

Rutherford, who believed at the time that the Thomson model was essentially correct, had his associates examine alpha particle scattering data to corroborate the model. A single Thomson model atom is expected to produce a very small deflection of an alpha particle passing through it. Because of its net zero charge such an atom has no effect on the alpha particle until the alpha is inside the atom. Once inside, the alpha interacts with the electrons and the positive charge. The electrons, however, have such a small mass compared to the alphas that the alphas are hardly deflected at all by them— much as a bowling ball is unaffected by an inflated rubber beach ball in its path. And the positive charge, because it is distributed over the volume of the Thomson atom, cannot produce a large deflection on an alpha particle either. One can make an estimate (see Prob. 4) that the average deflection of an alpha caused by one Thomson atom (see Fig. 7-3a) is not more than about 10^{-4} radian. After passage through many atoms in the foil, the cumulative effect of the deflections is still not large because of the randomness of the encounters—the alpha is deflected a small amount this way and then a small amount that way (see Fig. 7-3b). One can use statistical theory to show that, for the thin foils used, much less than 1% of the alphas should be scattered at angles greater than 3° and that the chance of a scattering of 90° or more (a backward scattering) was about one in 10^{3500}. The Thomson model atom involves small angle scattering from *many* atoms.

The results of the initial experiments confirmed the predominance of small atom scattering but the percentage of particles scattered at large angles was in disagreement with the predictions of the Thomson model. Indeed, one alpha in ten thousand (1 in 10^4, *not* 1 in 10^{3500}) came off backwards! To scientists accustomed to thinking in terms of Thomson's model it came as a great surprise that some α-particles were deflected through very large angles,

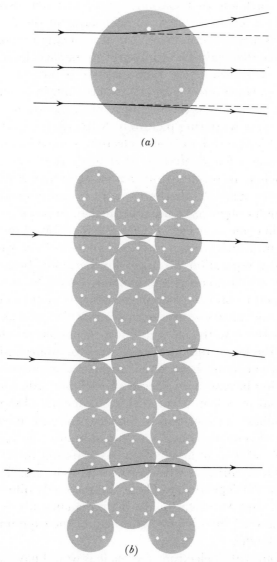

(a)

(b)

Figure 7-3. (a) Showing the deflection (greatly exaggerated) of various α-particles in passing through the single Thomson atom. (b) Showing the deflection (greatly exaggerated) of various α-particles in passing through a foil containing many layers of Thomson model atoms. The Thomson model atom involves small angle scattering from many atoms.

up to 180°. In Rutherford's words (see Ref. 3), "It was quite the most incredible event that ever happened to me in my life. It was as incredible as if you fired a 15-inch shell at a piece of tissue paper and it came back and hit you. On consideration I realized that the scattering backwards must be the result of a single collision, and when I made the calculation I saw that it was impossible to get anything of that order of magnitude unless you took a system in which the greater part of the mass of the atom was concentrated in a minute nucleus. It was then that I had the idea of an atom with a minute massive center carrying a charge."

Alpha particles traverse a different number of atoms N in passing through foils of different thickness. In Thomson's multiple scattering model, the alpha particle traverses a Thomson atom in each layer of the foil (see Fig. 7-3b) but its deflections are not cumulative, the alpha being randomly scattered—now this way, then that way. The number of alpha particles scattered through any angle can be shown to be proportional to the square root of N in Thomson's model. However, subsequent experiments in which foils of various thicknesses were used showed this number to be proportional to N itself. This is consistent with the scattering picture in Rutherford's model (see Fig. 7-4). The scattering of alphas due to atomic electrons can be ignored for scattering angles greater than a few degrees, so that Rutherford concentrated on the effect of the nucleus in seeking to explain large angle scatterings. Because the nucleus is very small, Rutherford's atom is mostly empty space so that alpha particles don't often come near the nucleus in passing through the foil. But those that do come near feel very large forces, for all the positive charge is concentrated in a small region and the Coulomb force varies inversely as the square of the separation distance. Furthermore, because the nucleus contains almost all the mass of the atom, it would be the (lighter) alpha particle that would recoil now. In other words, the large scale scatterings are caused by near encounters with nuclei whereas the predominance of small angle scatterings is due to the fact that most encounters are distant ones. The thicker the foil, the larger the number of layers of atoms and the greater the chance of a near encounter between an alpha and an atom. Hence, for thin foils, the scattering beyond a few degrees in Rutherford's model is expected to be proportional to N, consistent with experimental observations.

One can calculate exact scattering formulas for each model of the atom. Apart from the dependence on N mentioned above, the Rutherford model predicts that the relative number of alphas scattered to an angle ϕ should be proportional to $1/\sin^4 \phi/2$, whereas the Thomson model prediction gives an exponential ($e^{-\phi^2}$) dependence. These become vastly different as the angle increases. Geiger and Marsden, in Rutherford's laboratories, performed experiments that showed the striking agreement of the experimental data with the Rutherford model.

In a target atom with Z electrons the nucleus would have a positive charge Ze whose magnitude influences the scattering distribution in Rutherford's formula. The value of Z was not known for different atoms but now the scattering results could be used to determine Z. The experimental result was that Z—which gives the nuclear charge number and the number of planetary electrons in a neutral atom—equaled the chemical atomic number of the atom in the periodic table. This conclusion was independently confirmed by subsequent experiments on the scattering of X rays from atoms. So completely did experiments confirm the Rutherford nuclear model of the atom that in spite of unresolved difficulties about the stability of the atom, it was adopted universally.

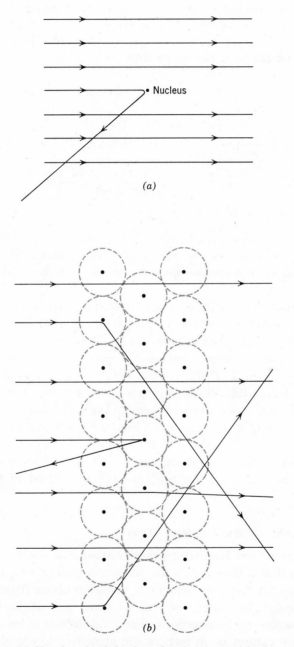

Figure 7-4. (*a*) Showing the deflection of various α-particles in passing the nucleus of a single Rutherford atom. (*b*) Showing the deflection of various α-particles in passing through a foil containing many layers of Rutherford model atoms. The Rutherford model atom involves large angle scattering from a single atom.

◆ **Example 2.** Rutherford had assumed that the incident α-particle was repulsed by the charged nucleus without penetrating it. The scattering of a beam of α's of speed 2.1×10^7 m/sec off of a gold foil conformed to the Rutherford scattering formula. The charge-to-mass ratio of an α-particle was known to be about 4.8×10^7 coul/kg. Take $Z = 79$ for gold atoms and estimate the closest distance of approach of an α to a gold nucleus.

The closest approach would correspond to a head-on "collision" in which the α-particle is scattered through 180°. The initial kinetic energy $\frac{1}{2} M_\alpha v_\alpha^2$ of the α is converted into mutual Coulomb potential energy $(1/4\pi\varepsilon_0)[(Ze)(2e)/D]$ as the α is brought to rest a distance D from the center of the nucleus, before being repulsed back along its original path. Hence, we have

$$\frac{1}{4\pi\varepsilon_0}\frac{(Ze)(2e)}{D} = \frac{1}{2}M_\alpha v_\alpha^2$$

so that

$$D = \left(\frac{2}{4\pi\varepsilon_0}\right)\frac{(Ze)(2e/M_\alpha)}{v_\alpha^2} = (2 \times 9.0 \times 10^9 \text{ nt-m}^2/\text{coul}^2)$$

$$\times \frac{(79 \times 1.6 \times 10^{-19} \text{ coul})(4.8 \times 10^7 \text{ coul/kg})}{(2.1 \times 10^7 \text{ m/sec})^2}$$

$$= 2.7 \times 10^{-14} \text{ m}.$$

This center-to-center distance of closest approach is about 10,000 times smaller than the assumed diameter of an atom. The nuclear volume then is about 10^{-12} that of the volume of an atom, suggesting the view that the atom is mostly empty space. ◀

It is found that the distribution of scattered α-particles at large scattering angles deviates from the predictions of the Rutherford formula as the incident α-particle energy is increased to large values. This is interpreted as being due to the fact that very energetic alphas can penetrate the nucleus. Then the scattering results should be affected not only by the Coulomb force between alpha and nucleus, which Rutherford took into account, but also by specifically nuclear forces, which were not known. We can define the nuclear radius as being the range of nuclear forces. Then, the nuclear radius can be determined from the value of the closest distance of approach for a head-on collision at that incident α energy at which deviations from Rutherford scattering set in. For gold, the nuclear radius is found in this way to be 6.9×10^{-15} m.

7.2 The Stability of the Nuclear Atom

At the center of the Rutherford model atom is a nucleus whose mass is approximately that of the entire atom and whose charge is equal to the atomic number Z times e; around the nucleus, at distances about 10,000 times greater than the nuclear radius, there exist Z electrons, neutralizing the atom as a whole. This picture raises questions about the *stability* of an atom.

The electrons cannot be at rest for the attractive Coulomb nuclear force would pull them into the nucleus, returning us to a nuclear-sized plum pudding model of the atom. One is then led to consider a planetary model in which electrons revolve about a nuclear "sun" in circular (or elliptical) orbits whose radii are of atomic size. Although a gravitational system such as this is stable, an electrical one is not. For, according to Maxwell, accelerated charged particles radiate electromagnetic waves. Electrons in planetary orbits are continually accelerating and should therefore be continually radiating energy. Hence, electrons would lose energy and spiral in toward the nucleus, just as a satellite in the atmosphere falls toward earth. The time during which an atom of diameter 10^{-10} m would collapse to nuclear size can be computed to be about 10^{-12} secs! Apart from contradicting the stability of atoms of size 10^{-10} m, this picture requires the emission of a continuous spectrum by the spiralling electrons, whereas atoms were known to emit discrete spectra. In fact, one reason Thomson sought a model with stationary electrons, that oscillate only on being disturbed, was to avoid the collapse that seemed inherent in a planetary model.

It was Neils Bohr who, in 1913, advanced a theory that preserved the stability of the nuclear atom and predicted a discrete atomic emission spectrum. We now examine what was then known about atomic spectra, preparatory to discussing the Bohr theory.

7.3 The Spectra of Atoms

It was known for some time that every atom had its own characteristic spectrum. Whereas solids or liquids emit a continuous spectrum, which is easily observed when they are at high temperatures, individual free atoms are observed to emit a spectrum consisting of discrete wavelengths. The spectrum can be produced by energizing atoms in a gas container, such as by passing an electric discharge through it. The atoms thereafter give up the energy they absorbed through collisions by emitting radiation and thereby return to their lowest energy state. In Fig. 7-5 we show schematically how the radiation from a source is made to reveal its spectrum. After being collimated by a slit system, the radiation passes through a prism and each wavelength in the spectrum is separately recorded on a photographic plate. The plate reveals a series of lines, each one an image of the line slit formed by a different wavelength.

In Fig. 7-6 we show the spectral lines emitted by hydrogen that span the visible and the near ultraviolet regions of the spectrum. The hydrogen atom is the simplest of all atoms and certainly must be understood if we are to understand the more complicated ones. Furthermore, most of the universe

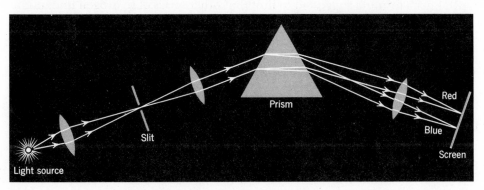

Figure 7-5. A prism spectrometer.

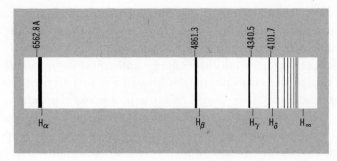

Figure 7-6. Emission spectrum of the hydrogen atom in the visible and the near ultraviolet region. H_∞ gives the theoretical position of the series limit. From Herzberg, *Atomic Spectra,* 1944.

consists of isolated hydrogen atoms so that its spectrum is of much practical interest, too. Note, in Fig. 7-6, that the hydrogen lines seem to form a converging series, the spacing between lines decreasing steadily towards a limit in the ultraviolet.

In 1884 Johann Balmer found that he could express by an empirical formula the frequencies of the fourteen then-known hydrogen lines. Put later by J. R. Rydberg in terms of reciprocal wavelengths (frequency divided by c, the speed of light) Balmer's formula became

$$\frac{1}{\lambda} = R_{\mathrm{H}}\left(\frac{1}{2^2} - \frac{1}{n^2}\right) \qquad n = 3, 4, 5, \ldots . \qquad (7\text{-}1)$$

Here R_{H} is a constant, now called the Rydberg constant, and n takes on all consecutive integral values greater than two. The Rydberg constant for hydrogen is known today to have the value $R_{\mathrm{H}} = 10967757.6 \pm 1.2$ m^{-1}, which suggests how great the precision of spectral measurements can be. The set of lines given by Balmer's formula is now called the Balmer series (see Ref. 4). By the time Rutherford's model was accepted, over thirty lines of the Balmer series had been found, all having wavelengths predicted by his formula. Balmer correctly concluded that there was a spectral *series limit,* the shortest wavelength (3638 Å), corresponding to $n = \infty$ and that the first visible hydrogen line (6563 Å) was the longest possible of the series, with $n = 3$.

Balmer's formula suggested to him that other series of hydrogen lines might exist. He correctly inferred that his original formula could be generalized to

$$\frac{1}{\lambda} = R_{\mathrm{H}}\left(\frac{1}{m^2} - \frac{1}{n^2}\right) \qquad (7\text{-}2)$$

in which m takes on a fixed integral value and n takes on all integral values with $n \geq m + 1$. With $m = 2$ and $n \geq 3$, for example, we obtain Balmer's original formula. In the early 1900's new hydrogen lines were discovered in the far ultraviolet and in the infrared and subsequently Balmer's generalized formula was confirmed over a wide spectral range. In addition to the Balmer series, there is

the Lyman series, with $m = 1$ and $n \geq 2$, in the ultraviolet;
the Paschen series, with $m = 3$ and $n \geq 4$, in the near infrared;
the Brackett series, with $m = 4$ and $n \geq 5$, in the infrared; and
the Pfund series, with $m = 5$ and $n \geq 6$, in the far infrared.

◆ **Example 3.** From the generalized Balmer formula, calculate the wavelength range within which each of the five observed discrete series of hydrogen lines falls.

With $R_{\text{H}} = 10967757.6$ m^{-1} and 1 Å $= 10^{-10}$ m, we have

$$\textit{Lyman series} \qquad \frac{1}{\lambda_1} = R_{\text{H}}\left(\frac{1}{1^2} - \frac{1}{2^2}\right) \qquad \lambda_1 = 1216 \text{ Å}$$

$$\frac{1}{\lambda_\infty} = R_{\text{H}}\left(\frac{1}{1^2} - \frac{1}{\infty}\right) \qquad \lambda_\infty = 912 \text{ Å}$$

$$\textit{Balmer series} \qquad \frac{1}{\lambda_1} = R_{\text{H}}\left(\frac{1}{2^2} - \frac{1}{3^2}\right) \qquad \lambda_1 = 6563 \text{ Å}$$

$$\frac{1}{\lambda_\infty} = R_{\text{H}}\left(\frac{1}{2^2} - \frac{1}{\infty}\right) \qquad \lambda_\infty = 3640 \text{ Å}$$

$$\textit{Paschen series} \qquad \frac{1}{\lambda_1} = R_{\text{H}}\left(\frac{1}{3^2} - \frac{1}{4^2}\right) \qquad \lambda_1 = 18,760 \text{ Å}$$

$$\frac{1}{\lambda_\infty} = R_{\text{H}}\left(\frac{1}{3^2} - \frac{1}{\infty}\right) \qquad \lambda_\infty = 8210 \text{ Å}$$

$$\textit{Brackett series} \qquad \frac{1}{\lambda_1} = R_{\text{H}}\left(\frac{1}{4^2} - \frac{1}{5^2}\right) \qquad \lambda_1 = 40,530 \text{ Å}$$

$$\frac{1}{\lambda_\infty} = R_{\text{H}}\left(\frac{1}{4^2} - \frac{1}{\infty}\right) \qquad \lambda_\infty = 14,590 \text{ Å}$$

$$\textit{Pfund series} \qquad \frac{1}{\lambda_1} = R_{\text{H}}\left(\frac{1}{5^2} - \frac{1}{6^2}\right) \qquad \lambda_1 = 74,620 \text{ Å}$$

$$\frac{1}{\lambda_\infty} = R_{\text{H}}\left(\frac{1}{5^2} - \frac{1}{\infty}\right) \qquad \lambda_\infty = 22,800 \text{ Å}$$

In each case λ_∞ is called the *series limit*. ◆

The emission spectra of elements other than hydrogen were also observed to be series of lines and in the 1890's Rydberg found approximate empirical laws to describe some of them, such as the alkali elements. His formulas for reciprocal wavelengths contained for each element a constant and a difference of squares. These Rydberg constants, however, had very slightly larger values than the hydrogen one, the deviation from hydrogen increasing the heavier the element. Absorption spectra, too, were known, i.e., the spectrum of lines that are absorbed by atoms from a continuous spectrum in a beam incident upon them. We shall discuss their characteristics later. It is clear, however, that by 1913 a large amount of accurate spectroscopic data of various kinds was available, with much of it organized into empirical formulas, but for which no theoretical explanation existed. It was at this time that Neils Bohr was trying to construct an atomic theory by synthesizing Rutherford's model of the atom and Planck's quantization of the energy of oscillators. He later reported "As soon as I saw Balmer's formula, everything became clear to me."

7.4 The Bohr Atom

In 1913, Bohr (Ref. 5) presented his theory of the atom. His postulates were grounded in experimental results and, more so than Planck, he showed a willingness to give up those established classical ideas which simply contradicted observation. Thus he postulated first that atoms exist in "stationary states" of definite energy, in which states they do not radiate energy and are stable. If one pictures an electron orbiting about a nucleus in the hydrogen atom, as Bohr did, this postulate contradicts the classical electromagnetic assertion that the electron must radiate energy. This first Bohr postulate, however, is equivalent to the idea of *the quantization of energy of atoms*. One need not invoke a specific classical picture of orbiting electrons, or, indeed, *any* picture to accept this postulate, but we shall see later that there is a plausible and satisfying picture of the atom in a stationary state that does not do violence to the electromagnetic ideas.

Bohr next postulated that the atom emits or absorbs radiant energy *only* when the atom goes from one stationary state to another and that the change in energy of the atom equals the energy of the photon of radiation that is emitted or absorbed. That is, if the atom makes a transition between allowed energy states of energies E_2 and E_1, where E_2 is greater than E_1, then (*a*) if the transition is from (higher) E_2 to (lower) E_1 we have

$$E_2 = E_1 + h\nu_{21}$$

in which ν_{21} is the frequency of the *emitted* photon, whereas (*b*) if the transition is from (lower) E_1 to (higher) E_2 we have

$$E_1 + h\nu_{12} = E_2$$

in which ν_{12} is the frequency of the *absorbed* photon. This second postulate, called *the radiation condition,* contains the conservation of energy idea, of course, but more specifically invokes Einstein's quantum picture of radiation.

Finally, Bohr presented what can be called a quantization rule, that is, a way to determine the allowed energy states of the atom. This rule, now called the *correspondence principle,* states most generally that quantum theory must give the same results as classical theory in the limit in which classical theory is known to be correct. This idea has been used already in relativity where we showed that as $v/c \rightarrow 0$ the relativistic results become the same as the classical results, which are known to be valid in the region of low speeds. In the case of the hydrogen atom, the correspondence principle requires that in the limit of large systems, where the allowed energies form a continuum, the quantum radiation condition must yield the same result as a classical calculation. If one pictures an electron orbiting about a nucleus, this means that for *very* large orbital radii, such that the atom has macroscopic size, the frequency of the radiation emitted by hydrogen should be the same as the frequency of revolution of the electron. Let us now apply Bohr's postulates.

We shall proceed as Bohr did. First we will use the experimentally verified Balmer formula together with Bohr's first two postulates to determine the allowed energies of the hydrogen atom. After getting familiar with this result, we shall then *derive* it from Bohr's third postulate, that is, from theory alone independent of experimentally determined quantities.

We start with the Balmer formula $1/\lambda = R_{\mathrm{H}}(1/m^2 - 1/n^2)$ and, to get it in terms of emitted frequency ν_{nm}, multiply by c, yielding $\nu_{nm} = cR_{\mathrm{H}}(1/m^2 - 1/n^2)$. The energy of the radiated photon, $h\nu_{nm}$, is then

$$h\nu_{nm} = hcR_{\mathrm{H}}\left(\frac{1}{m^2} - \frac{1}{n^2}\right) \tag{7-3}$$

From Bohr's second postulate for emission, however, we have, for a transition from a stationary state of energy E_n to one of energy E_m

$$E_n - E_m = h\nu_{nm}. \tag{7-4}$$

Comparison of these equations shows that *the allowed energies of the stationary states* must be

$$E_n = -\frac{hcR_{\mathrm{H}}}{n^2} \tag{7-5}$$

in which each integer n determines an allowed (quantized) energy. The lowest (most negative) energy corresponds to $n = 1$ and the highest to $n = \infty$. The negative sign in Eq. 7-5 signifies that the system (hydrogen atom) is a *bound* one, whose total energy is negative. The binding is just barely broken when, on the planetary picture, enough energy is added to separate the electron and nucleus; they are then at rest an infinite distance apart, the total (kinetic plus potential) energy of the system being zero. If this energy, or more, is added, the atom is ionized.

▶ **Example 4.** Calculate the binding energy of the hydrogen atom, that is, the energy binding the electron to the nucleus.

The total energy of the highest bound state of a hydrogen atom is zero, obtained by putting $n = \infty$ into Eq. 7-5. The binding energy is therefore numerically equal to the energy of the lowest state, corresponding to $n = 1$ in Eq. 7-5. We obtain

$$
\begin{aligned}
E_1 &= -\frac{hcR_{\mathrm{H}}}{(1)^2} \\
&= -\frac{(6.63 \times 10^{-34}\text{ joule-sec})(3.00 \times 10^8\text{ m/sec})(1.1 \times 10^7\text{ m}^{-1})}{1} \\
&= -2.17 \times 10^{-18}\text{ joule} = -13.6\text{ ev}
\end{aligned}
$$

which agrees with the experimentally observed binding energy (13.6 ev) for hydrogen. ◀

In Fig. 7-7 we show an energy level diagram for hydrogen, with the quantum number n for each level. The energies shown are computed from Eq. 7-5. Notice that the levels become closer together as n increases and, from the fact that there are an infinite number of levels, we see that the classical continuum is approached at high values of n. Normally the atom is in the state of lowest energy, $n = 1$, called the ground state ("ground" state means "fundamental" state, the term originating from the German word "grund," meaning fundamental). When the atom absorbs energy from some excitation process it makes a transition to some higher allowed energy state, called an excited state, for which $n > 1$. The atom may emit this energy thereafter in a series of transitions to allowed states of lower energy and return to its ground

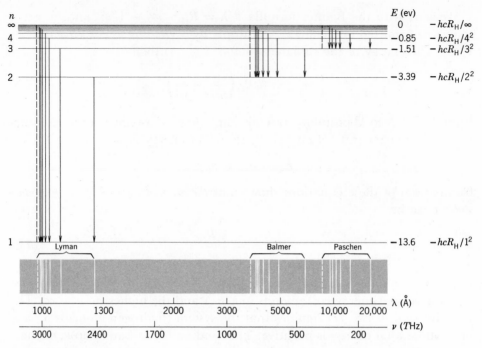

Figure 7-7. *Top:* the energy level diagram for hydrogen with the quantum number *n* for each level, the energy of the atom in each quantum state, and some of the transitions between states giving rise to radiation that appears in the spectrum. An infinite number of levels is crowded in between the levels marked $n = 4$ and $n = \infty$. *Bottom:* the corresponding spectral lines for the three series indicated above. Within each series the spectral lines follow a regular pattern, approaching the series limit at the short-wave end of the series. Note that as drawn here, neither the wavelength nor frequency scales are linear, being chosen as they were merely for clarity of illustration. The Brackett and Pfund series (not shown here) lie in the far infrared part of the spectrum.

state. For each transition a photon is emitted. In a spectral discharge tube containing a large number of atoms, all possible transitions occur and the entire spectrum is emitted. The spectrum can then be analyzed as consisting of distinct series of lines. The transitions between allowed stationary states of the hydrogen atom that give rise to the observed spectral series of lines are shown in the figure, the Balmer series, for example, arising from transitions that end on the second stationary state, $n = 2$. Actually, Bohr had predicted the existence of a stationary state corresponding to $n = 1$ before the Lyman series of lines, for transitions ending in this ground state, had been observed.

Up to now we have been examining and interpreting Eq. 7-5. But that equation is an empirical relation that combines Bohr's first two postulates—the quantization of energy and the radiation condition—with the generalized Balmer formula. In that formula the Rydberg constant is determined by experiment. Bohr used his third postulate to *derive* a relation giving the Rydberg constant in terms of other known fundamental constants. His theoretically determined value agreed with the precisely determined experimental value, which was a triumph for his theory. Let us now use the third postulate—the correspondence principle—to determine the Rydberg constant.

In the classical picture of the hydrogen atom an electron revolves about the nucleus with a frequency ν of revolution that is easily shown to be

$$\nu = \frac{4\varepsilon_0}{e^2}\sqrt{\frac{2}{m}}\,|E|^{3/2},\qquad(7\text{-}6)$$

in which E is the total (kinetic plus potential) energy of the system. (The algebraic manipulation needed to get this result is diverting, so we leave the proof—which is quite simple in principle—to a guided problem, Prob. 18). The accelerating electron should radiate continuously at a frequency equal to its frequency of revolution; as it loses energy and spirals inward it radiates at continuously higher frequencies. In the quantum picture, however, the atom—starting in a state of high energy—jumps from one stationary state to another lower one, emitting a photon in each transition, which gives rise to a discrete set of spectral lines. At high energies, however, the allowed states are very close together in energy (see Fig. 7-7) so that the successively emitted photons should be very close together in frequency. This is the region in which the quantum levels approach a classical continuum and in which, according to the correspondence principle, the quantum frequency condition should yield the known classical result.

Imagine an atom passing successively through every stationary state. For a transition from state n to state $n-1$ the quantum radiation condition gives us

$$\nu = cR_{\mathrm{H}}\left[\frac{1}{(n-1)^2}-\frac{1}{n^2}\right]$$
$$= cR_{\mathrm{H}}\left[\frac{2n-1}{(n-1)^2 n^2}\right].$$

At very large values of n we can neglect 1 in comparison to n so that this formula becomes

$$\nu = cR_{\mathrm{H}}\frac{2n}{n^2 n^2}=\frac{2cR_{\mathrm{H}}}{n^3}.\qquad(7\text{-}7)$$

This is the quantum result in the "classical" region of large quantum numbers—the region where the energy levels are so close together as to form a continuum for practical purposes—and, according to the correspondence principle, it should agree with the classical result, which assumes that the allowed energies *do* form a continuum. The classical result, Eq. 7-6, when combined with Eq. 7-5 for the total energy, $E_n = -hcR_{\mathrm{H}}/n^2$, is

$$\nu = \frac{4\varepsilon_0}{e^2}\sqrt{\frac{2}{m}}\frac{(hcR_{\mathrm{H}})^{3/2}}{n^3}.\qquad(7\text{-}8)$$

Equating Eqs. 7-7 and 7-8 and solving for the Rydberg constant for hydrogen we obtain

$$R_{\mathrm{H}}=\frac{e^4 m}{8\varepsilon_0^2 h^3 c}\qquad(7\text{-}9)$$

Hence, we now have a theoretically predicted value for the Rydberg constant in terms of other fundamental constants—the charge e and mass m of an electron, the speed c of light, and Planck's constant h.

Bohr, using data available in his time for these fundamental constants, obtained good agreement with experiment, the agreement today being within the extremely narrow limits of experimental error. Therefore, we can now regard the constant R_H as theoretically determined and write the allowed energies of the hydrogen atom, Eq. 7-5, in terms of other constants as

$$E_n = -\left(\frac{me^4}{8\varepsilon_0^2 h^2}\right)\frac{1}{n^2}. \tag{7-10}$$

In summary, then, Bohr's postulates enabled him to deduce a theoretical formula for the allowed energies of the hydrogen atom and to establish rules for emission of radiation that agreed with the observed wavelengths in the hydrogen emission spectrum. The only place that a classical picture was invoked—the picture of a point electron orbiting in a definite circular path about a nucleus—was in the classical limit where such a picture is valid according to the correspondence principle.

7.5   The Bohr Model of the One-Electron Atom

The Bohr postulates met with other successes. It was possible, for example, to understand absorption spectra in terms of the Bohr atom. A Bohr atom can absorb only certain energies from a beam of incident radiation. If a continuous spectrum of wavelengths is incident on a container of gas, the radiation transmitted is found to be missing a set of discrete wavelengths that must have been absorbed by the atoms in the container. Each so-called absorption line has the same wavelength as a line in the emission spectrum of the atom. However, not every emission line appears in the absorption spectrum. This is explained by noting that the atoms in a gas are normally in the ground state. Hence in hydrogen, for example, only those lines corresponding to exciting the atom from the ground state ($n = 1$) will appear in the absorption spectrum and indeed, only the Lyman series is normally observed. If the gas is at a high temperature, many of the atoms may be in excited states to begin with so that absorption lines corresponding to the Balmer series, for example, may appear. This, in fact, is observed to be the case in stellar spectra (see Quest. 19).

With his postulates confirmed, Bohr then ventured further. Rather than limiting himself to correspondence between classical and quantum motion at large values of n, he constructed a model for the one-electron atom that pictured the electron as moving classically in *all* stationary states of the atom. This model was meant to be only suggestive, and, in fact, the specifics of the model did not survive in the final quantum mechanical picture that did emerge. Nevertheless, Bohr's semiclassical planetary model made plausible many observed properties of atoms and introduced key new ideas that did survive and that helped to develop the new theory. We now examine features of Bohr's model.

Consider an electron, charge e, moving about a nucleus of charge Ze in a circular orbit. From $F = ma$, we have

$$\frac{1}{4\pi\varepsilon_0} \frac{Ze^2}{r^2} = \frac{mv^2}{r} \tag{7-11}$$

so that the kinetic energy K is

$$K = \frac{1}{2}mv^2 = \frac{1}{8\pi\varepsilon_0} \frac{Ze^2}{r}.$$

The mutual electric potential energy U, (see *Physics*, Part II, Sec. 29-6) is

$$U = -\frac{1}{4\pi\varepsilon_0} \frac{Ze^2}{r}$$

so that the total energy $E = K + U$ is

$$E = -\frac{1}{4\pi\varepsilon_0} \frac{Ze^2}{2r}. \tag{7-12}$$

Bohr adopted this classical planetary picture as the starting point in his model of the one-electron atom. Such a formula would correspond to a hydrogen atom if we set $Z = 1$. It would apply to a singly ionized helium atom, if we set $Z = 2$, and to a doubly ionized lithium atom with $Z = 3$, etc. It is the general one-electron atom classical relation.

The generalized Balmer-Rydberg empirical formula for one-electron atoms was assumed to be $1/\lambda = Z^2 R_H[1/m^2 - 1/n^2]$. This was confirmed for singly ionized helium lines observed in 1896 by Pickering (see Prob. 32), and in subsequent laboratory measurements on other one-electron atoms. For example, singly ionized helium atoms He^+ or doubly ionized lithium atoms Li^{++} can be formed by sending a high voltage discharge through a container of the normal gas. The so-called spark spectrum emitted by these ions is much simpler than the "arc" spectrum emitted by the normal atom and is easily identified. It is found that each frequency of He^+, for example, is almost precisely four times that of each corresponding hydrogen frequency, consistent with the formula above in which $Z^2 = 4$.

Bohr found that his entire earlier interpretation and analysis for hydrogen was consistent with a generalization to one-electron atoms of nuclear charge greater than one, if, in his hydrogen analysis, he simply replaced e^2 by the more general $(Ze)e$, to account for the force between a nucleus of charge Ze and a single electron. Hence, in Eq. 7-10 for the energy, e^4 is replaced by Z^2e^4, and our one-electron energy formula becomes

$$E_n = -Z^2\left(\frac{me^4}{8\varepsilon_0^2 h^2}\right)\frac{1}{n^2}. \tag{7-13}$$

Since, according to his earlier postulates, the atom can exist only in these energy states, Bohr then modified the classical planetary picture so that the radius of the electron's orbit in Eq. 7-12 is allowed to take on only certain values. If now we combine Eqs. 7-12 and 7-13 we can solve for these *allowed orbital radii*, obtaining

$$r_n = \frac{(4\pi\varepsilon_0)\hbar^2}{mZe^2}n^2 \qquad n = 1, 2, 3, \ldots \tag{7-14}$$

This can be written as $r_n = a_0 n^2$, in which $a_0 = (4\pi\varepsilon_0)\hbar^2/mZe^2$ gives the first ($n = 1$) allowed Bohr radius. For hydrogen, with $Z = 1$, we obtain $a_0 = 5.3 \times 10^{-11}$ m $\simeq 0.5$ Å. This agrees in magnitude with what was known about the size of a hydrogen atom in its normal state. The orbits of the successively higher excited states in this picture are then $4a_0$, $9a_0$, $16a_0$, etc., so that a highly excited atom with a very large value of n can be regarded as approaching a classical macroscopic atom. In interstellar space, where the density of atoms is low and the mean free path between collisions is large, it is possible for hydrogen atoms to exist in such highly excited states of large orbital radius with little disturbance and (see Quest. 19) emissions from states of high n are characteristically observed there rather than in high density discharge tubes.

Corresponding to each allowed orbital radius is an allowed orbital velocity. From Eq. 7-11, we have $v^2 = (1/4\pi\varepsilon_0)(Ze^2/mr)$ and, on substituting for r the allowed orbital radii r_n of Eq. 7-14, we obtain the *allowed orbital velocities*

$$v_n = \frac{Ze^2}{(4\pi\varepsilon_0)\hbar}\frac{1}{n} \qquad n = 1, 2, 3, \ldots \qquad (7\text{-}15)$$

We find the highest velocity to be in the ground state, $n = 1$, and for hydrogen, with $Z = 1$, its value in terms of the speed of light is

$$\frac{v_1}{c} = \left(\frac{1}{4\pi\varepsilon_0}\right)\frac{e^2}{\hbar c} = \frac{1}{137}.$$

Since, in this picture, the velocity of the electron decreases as the orbital radius increases ($v_n \propto 1/n$ in Eq. 7-15), this result justifies the use of non-relativistic mechanics for the hydrogen orbits. Of course, in one-electron atoms with large nuclear charge Z relativistic considerations become more important, the electron moving at higher speed in the stronger Coulomb field of such nuclei. Even in hydrogen there are small relativistic effects and such effects are taken into account in more detailed theories to explain the fine structure of spectral lines found in all atomic spectra.

The Bohr picture helps us to understand yet another feature of atomic spectra. We have seen (Example 4) that the binding energy, i.e., the energy difference between the ground state of an atom and its state of zero energy, is the minimum energy required to ionize a normal atom. In Bohr's orbiting planetary picture of a one-electron atom, the electron in an ionized atom would no longer be bound to the nucleus. The highest discrete bound energy state of the atom is $E_\infty = 0$. For positive total energies, the atom would be ionized, the electron no longer being bound and becoming a free particle. The electron's "orbit" is one of infinite radius, and the correspondence principle tells us that this is the classical region. Therefore, the energy of a free particle is not quantized, so that a continuum of energy exists above the highest quantized state at $E = 0$. If an energy greater than the binding energy is supplied to a one-electron atom, the electron will be free in this energy continuum. When a photon supplies this energy to the atom, we simply have the photoelectric effect. Correspondingly, an ionized atom can capture a free electron, the neutral atom thereafter being in an allowed quantized energy state. Radiation of frequency greater than that of the series limit for that state will be emitted in such a process. Since the free electron can have any energy $E > 0$ initially, there ought to be a continuum in the spectrum of the atom beyond each series limit. This, in fact, is observed experimentally under the appropriate circumstances.

◆ **Example 5.** (a) Find the allowed frequencies of revolution of an electron in the Bohr model one-electron atom.

If ω represents angular velocity, we can write Eq. 7-11 as $m\omega^2 r = Ze^2/4\pi\varepsilon_0 r^2$ or

$$\omega^2 = \frac{1}{4\pi\varepsilon} \frac{Ze^2}{mr^3}.$$

From Eq. 7-14, however, only certain radii, $r_n = \varepsilon_0 h^2/\pi m Z e^2 n^2$, are allowed. Substituting these values for r into the above equation and solving for ω yields

$$\omega_n = \frac{Z^2 m e^4}{8\varepsilon_0^2 h^2 n^2} \frac{4\pi}{hn},$$

as the student can verify. But $\omega = 2\pi\nu$, so that the allowed frequencies ν_n are

$$\nu_n = \frac{Z^2 m e^4}{8\varepsilon_0^2 h^2 n^2} \frac{2}{hn} = \frac{2|E_n|}{hn}$$

where E_n is given by Eq. 7-13.

(b) Compare the frequencies of revolution of an electron in an upper and lower state of the Bohr model atom to the frequency of radiation emitted in a transition between states.

We can write the frequencies of revolution (above) on Bohr's model as

$$\nu_n = \left(\frac{Z^2 m_e e^4}{8\varepsilon_0^2 h^3}\right) \frac{2}{n^3} \quad \text{and} \quad \nu_m = \left(\frac{Z^2 m_e e^4}{8\varepsilon_0^2 h^3}\right) \frac{2}{m^3}$$

for the upper nth and lower mth states. However, the frequency of radiation ν_{nm} emitted in a transition $n \to m$ is given by the radiation condition as

$$\nu_{nm} = \frac{E_n - E_m}{h} = \left(\frac{Z^2 m_e e^4}{8\varepsilon_0^2 h^3}\right)\left(\frac{1}{m^2} - \frac{1}{n^2}\right).$$

This observed frequency of radiation does *not* correspond to *either* of the previous frequencies. Indeed, one can show (see Prob. 22) that

$$\frac{2}{m^3} \geq \left(\frac{1}{m^2} - \frac{1}{n^2}\right) \geq \frac{2}{n^3}$$

so that the frequency of the emitted radiation lies *between* the frequencies of revolution of the electron in the orbits between which, in Bohr's picture, the transition occurs.

We would expect the Bohr picture to be strictly correct in the classical limit, and, indeed, if $n = m + 1$ and n is very large the radiation frequency equals the rotational frequency—the very condition Bohr originally imposed in his theory. But here again we see that in the quantum region, the classical picture must be discarded and the Bohr postulates used instead. ◆

7.6 Quantization of Angular Momentum

After Bohr had presented his model of the hydrogen atom, he sought a more direct way to calculate the allowed orbital radii. He noticed a remarkably simple result when he evaluated the orbital angular momenta, L, of the electron in its allowed circular orbits. The angular momentum is the moment of momentum, i.e., the linear momentum times the moment arm. Thus, from Eqs. 7-14 and 7-15,

$$L_n = mv_n r_n = m\left(\frac{Ze^2}{4\pi\varepsilon_0 \hbar}\frac{1}{n}\right)\left(\frac{4\pi\varepsilon_0 \hbar^2}{mZe^2}n^2\right)$$

or
$$L_n = n\frac{h}{2\pi} = n\hbar \qquad n = 1, 2, 3, \ldots \tag{7-16}$$

The orbital angular momentum of the electron is quantized, taking on only integral multiples of $\hbar$. This result was so simple that Bohr felt compelled to change his third postulate. As Bohr arrived at it, the quantization of the angular momentum was a result of his earlier postulates and his planetary model. But, with his intuition for simple central principles, Bohr made the quantization of angular momentum, rather than the correspondence principle, the central postulate for determining the quantized energy states of the atom. Starting from Eq. 7-16, one can derive all the results of the Bohr atom and the Bohr model of the one-electron atom (see Prob. 24) and can show thereafter that the results are consistent with the correspondence principle.

Bohr's original deduction of the quantized energy states of the one-electron atom (Sec. 7.4) did *not* require the pictorialization of a planetary atom. Since such a picture does not correspond to the modern quantum view, some prefer not to use it and therefore prefer Bohr's original approach. Bohr's subsequent deduction (Secs. 7-5 and 7-6) *does* require a planetary model, but the idea of the quantization of angular momentum that he used in this approach proved to be basic to modern quantum theory.

In 1924 de Broglie gave a physical interpretation of the Bohr quantization rule for angular momentum. If p represents the linear momentum of an electron moving in an allowed circular orbit, we can write the angular momentum as pr and Eq. 7-16 becomes

$$pr = n\frac{h}{2\pi} \qquad n = 1, 2, 3, \ldots$$

But, in terms of the de Broglie wavelength, we can write $p = h/\lambda$ and the equation for angular momentum becomes

$$\frac{h}{\lambda}r = n\frac{h}{2\pi}$$

or
$$2\pi r = n\lambda \qquad n = 1, 2, 3, \ldots \tag{7-17}$$

Therefore, only those orbits are allowed in which the circumference contains an integral number of de Broglie wavelengths.

Imagine the electron to be moving in a circular orbit with constant speed and to have a de Broglie wave associated with it. The wave, of wavelength λ, is then wrapped repeatedly around the circular orbit. The resultant wave that is produced will have zero intensity at any point unless the wave at each passage is exactly in phase at that point with the wave in other passages. If the waves in each passage are exactly in phase, then the orbits must contain

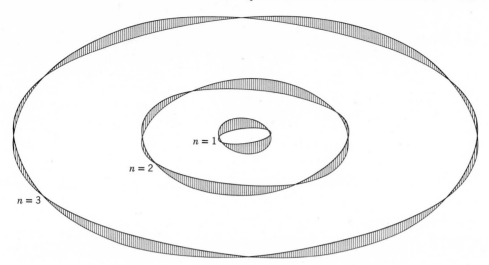

Figure 7-8. A schematic representation of the envelopes of standing de Broglie waves set up in the first three Bohr orbits. The location of the nodes is, of course, arbitrary.

an integral number of de Broglie wavelengths, as required by Eq. 7-17. If this requirement were not met, then in a large number of passages the waves would interfere with one another such that their average intensity would be zero. However, the average intensity of the waves, $\overline{\Psi}^2$, gives the probability of locating the particle, so that an electron has no chance of being in an orbit in which $\overline{\Psi}^2$ is zero.

This wave picture gives no suggestion of progressive motion. Rather, it suggests standing waves, as in a stretched string of a given length. In a stretched string only certain wavelengths, or frequencies of vibration, are permitted. Once such modes are excited, the vibration goes on indefinitely if there is no damping. To get standing waves, however, we need oppositely directed traveling waves of equal amplitude. For atoms this requirement is presumably satisfied by the fact that the electron can traverse an orbit in either direction and still have the magnitude of angular momentum required by Bohr. The de Broglie standing wave interpretation, illustrated in Fig. 7-8, therefore provides a satisfying basis for Bohr's quantization condition.

Subsequently, the Bohr picture for hydrogen had to be abandoned. Because of the uncertainty principle, electron orbits cannot even be precisely defined quantum mechanically. Furthermore, in the case of hydrogen the lowest state has zero angular momentum rather than the value $\hbar$ predicted by Bohr. Nevertheless, orbital angular momentum *is* quantized in quantum mechanics and takes on only values of zero or integral multiples of $\hbar$. Moreover, the de Broglie standing wave idea is essentially retained, in this manner. Instead of locating the electron in an atom precisely, we speak of the (three-dimensional) probability per unit volume Ψ_n^2 of finding the electron in an atom in quantum state n. The distribution of charge that is obtained when we multiply this probability function by the charge e of the electron is called an electronic cloud surrounding the nucleus. The absence of radiation for stationary states—which was simply postulated in the Bohr theory—comes about by the fact that the elements of radiation, emitted in the classical picture by the individual moving elements of the electronic cloud, annul each other by interference. We have a standing wave configuration whose energy content does not vary with time. Furthermore, the Bohr radiation condition, $h\nu_{nm} = E_n - E_m$, can be derived by application of the correspondence principle, and emerges as a beat frequency between the two corresponding stationary states of the atom.

7.7   Correction for the Nuclear Mass

So far in our analysis of the Bohr model we have assumed that the nucleus remains at rest as the electron revolves about it. This is equivalent mechanically to regarding the nucleus as having an infinite mass compared to the electron. Because the mass of the hydrogen atom nucleus is nearly 2000 times larger than the mass of the electron, our procedure has been approximately correct. However, just as two bodies of finite mass which are under the influence of each other's gravitational attraction move in circular orbits about their common center of mass with the same angular frequency (see *Physics,* Part 1, Sec. 16.7), so here the electron and nucleus move similarly under the influence of each other's electrical attraction. Because spectral data can be determined to such high accuracy, it turns out to be necessary, in order to get agreement with the data, to take into account in our formulas the actual finite mass of the nucleus and its effect on the motion. This can be done rather easily because in such a planetarylike system the electron moves relative to the nucleus as though the nucleus were fixed and the mass of the electron m were reduced to μ, the reduced mass of the system. The equations of motion of the system are the same as those we have considered if we simply substitute μ for m, where

$$\mu = m\left(\frac{M}{m + M}\right), \tag{7-18a}$$

in which M is the mass of the nucleus (see *Physics,* Part 1, Sec. 15.8). Notice that μ is *less* than m by a factor $1/(1 + m/M)$.

In Bohr's planetary model treatment of the one-electron atom he made the necessary correction by taking into account the angular momentum of the nucleus as well as that of the electron. He simply postulated that *the total orbital angular momentum of the whole atom* (not just the electron) *is an integral multiple of* $\hbar$, so that Eq. 7-16 is generalized to

$$\mu v r = n\hbar \qquad n = 1, 2, 3, \ldots \tag{7-18b}$$

If now we were to proceed with the Bohr analysis we would find the equations to be the same as before except that we must replace the electronic mass by the reduced mass, (see Prob. 27). In particular, the Rydberg constant for finite nuclear mass R_M is related to the one derived by Bohr (Eq. 7-9) for infinite nuclear mass, $R_\infty \equiv me^4/8\varepsilon_0^2 h^3 c$, by

$$R_M = \frac{\mu e^4}{8\varepsilon_0^2 h^3 c} = \frac{\mu}{m} R_\infty = \left(\frac{M}{m + M}\right) R_\infty \qquad (7\text{-}19)$$

The experimental value of R_H cited in Sec. 7-3 should be compared to this value R_M of the theory, rather than to the value $R_\infty = 109737$ cm^{-1} which incorrectly assumed infinite nuclear mass for hydrogen. When this is done, it is found that the values for hydrogen

$$\left[R_H = \left(\frac{M}{m + M}\right) R_\infty \simeq \left(\frac{1836}{1837}\right) R_\infty = 109678 \text{ cm}^{-1}\right]$$

agree to six significant figures! We see also in Eq. 7-19 confirmation of Rydberg's empirical findings mentioned earlier that the effective Rydberg constant increases slightly the heavier the atom. For example, $R_H = 109678$ cm^{-1}, $R_D = 109707$ cm^{-1}, and $R_{He^+} = 109722$ cm^{-1} give the Rydberg constants for hydrogen, deuterium (see Example 6), and singly ionized helium, respectively. Finally, with the correct Rydberg constant the formula for the reciprocal wavelength of the spectral lines becomes

$$\frac{1}{\lambda} = R_M Z^2 \left(\frac{1}{m^2} - \frac{1}{n^2}\right) \qquad (7\text{-}20)$$

with R_M, for an atom with a nucleus of mass M, being given by Eq. 7-19.

▶ **Example 6.** Ordinary hydrogen contains about one part in six thousand of *deuterium*, or heavy hydrogen. This is a hydrogen atom whose nucleus (containing a proton *and* a neutron) has a mass nearly twice that of a proton. How does this affect the observed spectrum?

The spectrum would be identical if it were not for the correction for finite nuclear mass. For a normal hydrogen atom the nuclear mass M is 1836 times the mass m_e of an electron so that (Eq. 7-19)

$$R_H = R_\infty \mu/m_e = \frac{R_\infty}{\left(1 + \frac{m_e}{M}\right)} = \frac{109737 \text{ cm}^{-1}}{\left(1 + \frac{1}{1836}\right)} = 109678 \text{ cm}^{-1}.$$

For an atom of heavy hydrogen, or deuterium, the nuclear mass is about doubled so that

$$R_D = R_\infty \mu/m_e = \frac{R_\infty}{\left(1 + \frac{m_e}{M}\right)} = \frac{109737 \text{ cm}^{-1}}{\left(1 + \frac{1}{2(1836)}\right)} = 109707 \text{ cm}^{-1}.$$

Hence, R_D is a bit larger than R_H, so that (see Eq. 7-20) the spectral lines of the deuterium atom are shifted to slightly shorter wavelengths compared to hydrogen.

Indeed, deuterium was discovered in 1932 by H. C. Urey following the observation of the spectral lines (see Fig. 7-9). By increasing the concentration of the heavy isotope above its normal value in a hydrogen discharge tube, one now can enhance the intensity of the deuterium lines which, ordinarily, are too weak to detect. One then readily observes pairs of hydrogen lines; the shorter wavelength members of the pair

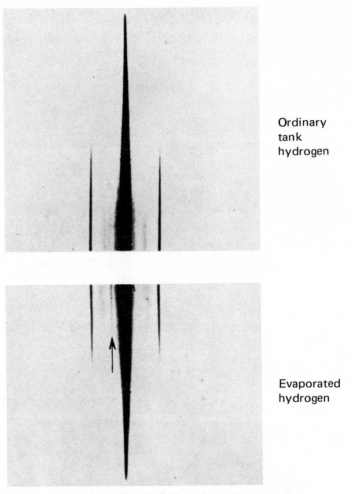

Ordinary
tank
hydrogen

Evaporated
hydrogen

Figure 7-9. The H-β lines for ordinary tank hydrogen (above) and for a sample of evaporated hydrogen (below). The outer lines are ghosts (due to periodicities in the grating spacings by the ruling engines). The main line in the center is due to hydrogen (H^1-β) and has about the same intensity for both exposures. The line to the left of the main line (see arrow) is the deuterium (H^2-β) line, which is considerably more intense in the lower case due to the increased concentration of H^2 in that sample. (From Urey, Brickwedde, and Murphy, *Physical Review,* **40,** No. 1, 1932.)

correspond exactly to those predicted from R_D above. The resolution needed is easily obtained, the H$_\alpha$-line pair being separated by about 1.8 Å, for example, several thousand times greater than the minimum resolvable separation.

Example 7. A *muonic* atom consists of a nucleus of charge Ze with a μ^--meson circulating about it. A μ^--meson is an elementary particle of charge $-e$ and a mass that is 207 times as large as an electron mass. Such an atom is formed when a proton, or some other nucleus, captures a μ^--meson.

(a) Calculate the radius of the first Bohr orbit of a muonic atom with $Z = 1$.

The reduced mass of the system, with $m_\mu = 207\ m_e$ and $M = 1836\ m_e$, is, from Eq. 7-18a,

$$\mu = \frac{(207\ m_e)(1836\ m_e)}{207\ m_e + 1836\ m_e} = 186\ m_e.$$

Then, from Eq. 7-14, with $n = 1$, $Z = 1$, and m replaced by $\mu = 186\ m_e$, we obtain

$$r_1 = \frac{(4\pi\varepsilon_0)\hbar^2}{186\ m_e e^2} = \frac{1}{186}(5.3 \times 10^{-11}\ \text{m}) = 2.8 \times 10^{-13}\ \text{m} = 2.8 \times 10^{-3}\ \text{Å}.$$

The μ^--meson then is much closer to the nuclear (proton) surface than is the electron in a hydrogen atom. It is this feature which makes such muonic atoms interesting, information about nuclear properties being revealed from their study.*

(b) Calculate the binding energy of a muonic atom with $Z = 1$. From Eq. 7-13, with $Z = 1$, $n = 1$, and $\mu(= m) = 186\ m_e$, we have

$$E_1 = -186\frac{m_e e^4}{8\varepsilon_0^2 h^2} = -(186)(13.6\ \text{ev}) = -2530\ \text{ev}$$

as the ground state energy. Hence, the binding energy is 2530 ev.

(c) What is the wavelength of the first line in the Lyman series for such an atom? From Eq. 7-20, with $Z = 1$, we have

$$\frac{1}{\lambda} = R_M\left(\frac{1}{m^2} - \frac{1}{n^2}\right)$$

For the first Lyman line, $n = 2$ and $m = 1$. In this case, $R_M = (\mu/m_e)R_\infty = 186\ R_\infty$. Hence,

$$\frac{1}{\lambda} = 186\ R_\infty\left(1 - \frac{1}{4}\right) = 139.5\ R_\infty.$$

With $R_\infty = 109737\ \text{cm}^{-1}$ we obtain

$$\lambda \simeq 6.5\ \text{Å}$$

so that the Lyman lines lie in the X-ray part of the spectrum. X-ray techniques are necessary therefore to study the spectrum of muonic atoms. ◀

7.8 The Quantization Conditions

The success of Bohr's theory of one-electron atoms encouraged others to apply his ideas to more complicated atoms and to other problems. In particular, Wilson and Sommerfeld in 1916 found a general rule that could explain Planck's quantization of the energy of an oscillator and Bohr's quantization of the angular momentum. Sommerfeld used this rule to work out a more detailed theory of the hydrogen atom, in which he considered three-dimensional elliptical orbits and relativistic effects of the electron's motion. He found that new energy levels, separated slightly from Bohr's, had to be introduced and that this fine structure of the levels was able to account for the double lines observed in highly resolved spectral photographs. Although the Wilson-Sommerfeld rule was not helpful in any but simple systems,

*See Ref. 6 for a discussion of muonic and other exotic atoms.

(see Probs. 36 and 37) we present it here both to show a connection between the Planck and Bohr quantization rules and because it foreshadowed concepts developed later in modern quantum theory.

Let q represent a coordinate in a periodic physical system and let p_q represent the momentum associated with that coordinate. Then, the Wilson-Sommerfeld rule states that

$$\oint p_q dq = n_q h \tag{7-21}$$

in which $\oint$ means that the integral is to be taken over a complete cycle of the periodic motion and n_q is an integral quantum number. For a one-dimensional harmonic oscillator, for example, the coordinate q is x and the corresponding momentum component is p_x. For circular motion the coordinate q is the angle θ and the corresponding momentum is the angular momentum L. Let us now apply the rule in each case.

Consider a one-dimensional simple harmonic oscillator. The total energy of oscillation can be written, in terms of position and momentum, as

$$E = K + U = \frac{p_x^2}{2m} + \frac{1}{2} kx^2$$

or

$$\frac{p_x^2}{2mE} + \frac{x^2}{(2E/k)} = 1.$$

The quantization integral $\oint p_x \, dx$ is most easily evaluated, for the relation between p_x and x given by this equation, if we consider its geometric interpretation. This is the equation of an ellipse. Any instantaneous state of the motion is represented by some point (x, p_x) in a plot of this equation on a two-dimensional space having coordinates p_x and x. We call such a space (the p-q plane) "phase space" and the plot is a phase diagram of the linear oscillator, shown in Fig. 7-10. During one cycle of oscillation the point representing the position and momentum of the particle travels once around the ellipse. The semi-axes a and b of the ellipse $p_x^2/a^2 + x^2/b^2 = 1$ are seen, by comparison with our equation, to be

$$a = \sqrt{2mE} \qquad b = \sqrt{\frac{2E}{k}}.$$

Now the area of an ellipse is πab, so

$$\oint p_x \, dx = \pi ab.$$

In our case

$$\oint p_x \, dx = 2\pi E \sqrt{\frac{m}{k}}.$$

But

$$\sqrt{\frac{k}{m}} = 2\pi\nu,$$

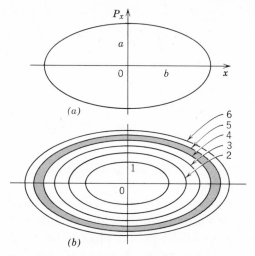

Fig. 7-10. (*a*) A diagram of the motion of a linear oscillator in phase space. (*b*) The allowed energy states of a linear oscillator are represented by ellipses whose areas in phase space are given by *nh*. The space between allowed states has an area *h* (as the shaded area).

where ν is the frequency of the oscillation, so that

$$\oint p_x \, dx = \frac{E}{\nu}.$$

If we now use Eq. 7-21, the Wilson-Sommerfeld equantization rule, we have

$$\oint p_x \, dx = \frac{E}{\nu} = n_x h \equiv nh$$

or

$$E = nh\nu$$

which is Planck's rule for the quantization of the energy of an oscillator, Eq. 4-12.

Note that the allowed states of oscillation are represented by a series of ellipses in phase space, the area enclosed between successive ellipses always being *h* (Fig. 7-10). We see again that the classical situation corresponds to $h \to 0$, all values of *E* and hence all ellipses being allowed there.

We can also deduce Bohr's rule for the quantization of angular momentum from the Wilson-Sommerfeld rule. An electron moving in a circular orbit of radius *r* has an angular momentum $mvr = L$, which is constant. The angular coordinate is θ, which is a periodic function of the time. That is, θ increases linearly from zero to 2π radians in one period and repeats this pattern in each succeeding period. Hence Eq. 7-21

$$\oint p_q \, dq = n_q h,$$

becomes

$$\oint L d\theta = nh.$$

But

$$\oint L d\theta = L \oint_0^{2\pi} d\theta = 2\pi L$$

so that

$$2\pi L = nh$$

or

$$L = \frac{nh}{2\pi} \equiv n\hbar$$

which is Bohr's rule for the quantization of angular momentum, Eq. 7-16.

We discussed earlier the more physical interpretation that de Broglie gave to Bohr's quantization of the angular momentum, namely that the waves associated with the electron in a circular orbit must form standing waves for stationary states of the atom. This standing wave condition is easily shown to be the same as the Wilson-Sommerfeld condition, Eq. 7-21. And, as indicated in Sec. 7-6, the standing wave concept makes plausible the absence of radiation and the stability of atoms in stationary states.

7.9 Conclusion

The Bohr theory and its extensions met with many successes, but their scope was clearly limited. Although applied successfully in many ways to one-electron atoms, and as an approximation to the alkali atoms, the theory could not even cope with the experimental observations on the neutral helium atom, let alone more complicated atoms. Even in the case of one-electron atoms, the theory had no prescription for calculating the *intensities* of the observed spectral lines. And, fundamentally more serious, neither Bohr's theory nor Planck's theory was a coherent theory of the physics of microscopic systems. Rather, these theories resembled patchwork in which some classical ideas which failed were declared not valid and replaced for certain special cases. What was needed was a reformulation and generalization of the laws of physics which would give the correct results for all systems—microscopic and macroscopic—reducing to the classical laws in the macroscopic domain. Such a reformulation was indeed made, starting with Schroedinger and Heisenberg in the late twenties, and the modern theory of quantum mechanics was born.*

Quantum mechanics is rather formal and abstract in comparison with earlier theories in physics. It is conceptually difficult for the mind trained to think at the level of macroscopic experience. Although familiarity with it soon enables one to feel at home with the theory, this is partly so because of the knowledge and experience gained with early quantum theory. The breakdown of classical ideas in the numerous experiments we have cited in the past four chapters, and the emergence of key ideas that explain our observations, not only motivate us to seek and accept a new theory but give

---

*See Ref. 7 for an interesting retrospective view of how a few roads not taken by physicists at the time might have made things happen faster than they did.

us an intuitive feeling and a conceptual basis for the theory that emerged. Indeed, every major idea we have discussed in these chapters remains valid and is incorporated into the new theory. This includes the quantization of energy of bound systems, the wave properties of matter and the particle properties of radiation, the wave-particle duality, the uncertainty principle and the probabilistic interpretation, the nuclear model of the atom, the correspondence principle, the quantization of the angular momentum, and the role of Planck's fundamental constant h. And throughout, experiment is the guide and test for theory.

Relativistic quantum mechanics can be applied to all atoms, to the nucleus, to molecules and solids and a quantum statistics can be developed that replaces the classical statistics. These topics are beyond the scope of this book and constitute a separate course. Here we have laid the necessary conceptual foundations that enable us to begin successfully such a study.

QUESTIONS

1. List objections to the Thomson model of the atom.

2. (*a*) Explain why scattering of alphas due to atomic electrons can be ignored for scattering angles greater than several degrees. (*b*) The scattering of α particles at very small angles disagrees with the Rutherford formula for such angles. Explain.

3. Explain why a head-on collision, $\phi = 180°$, gives the closest distance of approach in α-scattering from nuclei.

4. Explain why Rutherford could safely neglect the effect of the electrons in an atom in computing the distance of closest approach of an alpha particle to a nucleus.

5. How are chances of the α-scattering being due to only one nuclear encounter affected by increasing the thickness of the scattering foil? Why do we specify that the foil be thin in experiments intended to check the Rutherford scattering formula?

6. We have neglected the recoil of the target nucleus in calculating the distance of closest approach of an incident alpha. In principle, how could this be taken into account? In practice, does it matter for heavy nuclei?

7. Discuss the analogy between the Kepler-Newton relation in the development of the laws of gravitation and the Balmer-Bohr relation in the development of a theory of the atom.

8. Why was the Balmer series, rather than the Lyman of Paschen series, the first to be detected and analyzed in the hydrogen spectrum?

9. Did Bohr postulate the quantization of energy? What did he postulate?

10. The correspondence principle is expressed by taking n to infinity rather than taking h to zero, as we have usually done. Explain the relation between these two techniques for finding the classical limit.

11. On emitting a photon, the hydrogen atom recoils to conserve momentum. Explain the fact that the energy of the emitted photon is less than the energy difference between the energy levels involved in the emission process. (See Prob. 26.)

12. If only lines in the absorption spectrum of hydrogen need to be calculated, how would you modify Eq. 7-3 to obtain them?

13. For the Bohr hydrogen atom orbits, the potential energy is negative and greater in magnitude than the kinetic energy. What does this imply?

14. Why couldn't Bohr allow the quantum number n to take on the value $n = 0$, as it may in Planck's quantization equation? (*Hint*. see Eq. 7-15).

15. Keeping in mind the correspondence principle, what physical significance do you ascribe to Bohr orbits having very large values of n, say $n > 100$?

16. (*a*) Can a hydrogen atom absorb a photon whose energy exceeds its binding energy (13.6 ev)? (*b*) What minimum energy must a photon have to initiate the photoelectric effect in hydrogen gas? (Careful!)

17. Would you expect to observe all the lines of atomic hydrogen if such a gas were excited by electrons of energy 13.6 ev? Explain.

18. How would you estimate the temperature of hydrogen gas at which atomic collisions cause significant ionization of the atoms?

19. Only a relatively small number of Balmer lines can be observed from laboratory discharge tubes whereas a large number are observed in stellar spectra. Explain this in terms of the small density, high temperature, and large volume of gases in stellar atmospheres.

20. Classically all orbits are mechanically possible for an electron in a hydrogen atom, i.e., the atom can have any total energy whatsoever. Explain how this would lead to a continuous emission or absorption spectrum, even if one ignores radiation from accelerated charges.

21. According to classical mechanics, an electron moving in an orbit should be able to do so with any angular momentum whatever. According to Bohr's theory of the hydrogen atom, however, the angular momentum is quantized at values $L = nh/2\pi$. Reconcile these two statements, using the correspondence principle.

22. In what two ways does the Balmer formula for He^+ differ from that for neutral hydrogen? What effect does each difference have on the He^+ spectrum compared to hydrogen's?

23. Is the ionization energy of deuterium different from that of hydrogen? Explain.

24. Apply the correspondence principle to the phase diagram of a linear oscillator, Fig. 7-10. Explain.

PROBLEMS

1. If one assumes that the entire mass of the electron is electromagnetic in origin then one can equate the rest energy m_0c^2 to the energy in the electric field outside an electron, of charge e and assumed radius r, and finds that $r = 1/2(e^2/m_0c^2)$. Derive this result and show that $r \simeq 10^{-15}$ m.

2. Show, for a Thomson atom, that an electron moving in a stable circular orbit rotates at the same frequency with which it would oscillate in moving through the center along a diameter.

3. What radius must the Thomson model of a one-electron atom have if it is to radiate a spectral line of wavelength $\lambda = 6000$ Å? Comment.

★4. (*a*) An alpha particle of initial (nonrelativistic) velocity v collides with a free electron at rest. Show that, assuming the mass of the alpha particle to be about 7400 electronic masses, the maximum deflection of the alpha particle is about 10^{-4} radians. (*Hint*. Use momentum conservation). (*b*) Show that the maximum deflection of an alpha particle that interacts with the positive charge of a Thomson atom of radius 1.0 Å is also about 10^{-4} radian. (*Hint*. Estimate the maximum force and the time of interaction to determine the momentum transfer.) (c) In view of the results of (a) and (b), argue that ϕ_{max} is of the order of 10^{-4} radians for the scattering of an alpha particle by a single Thomson atom.

5. In the Thomson model the positive charge of an atom is distributed uniformly over a sphere of radius R about 10^{-10} m; in the Rutherford model . . . about 10^{-14} m. Calculate and plot the electric field from $r = 0$ to $r = R$, and from $r = R$ to $r = 10^{-9}$ m, produced by the positive charge on *each* model (see Part II, Sec. 28.6). Compare the results.

6. What is the distance of closest approach for a head-on collision of a 5.30 Mev α-particle to copper nuclei?

7. Assume the gold nucleus has a radius 7.0×10^{-15} m and the α-particle has a radius of 2.0×10^{-15} m. What minimum energy must an incident alpha have to experience non-Coulombic nuclear forces (i.e., to penetrate the nucleus)?

★8. When an alpha particle collides elastically with a nucleus, the nucleus recoils. (a) Use the fact that the motion of the center of mass of the system is unchanged, and find the recoil energy of a gold nucleus ($A = 197$) in terms of the initial kinetic energy K of the alpha. (b) Assume K is less than the energy needed for nuclear penetration and determine the distance of closest approach of the alpha. Compare this with the result one gets if, as in the text, we assume that the gold nucleus remains at rest.

9. Using Balmer's formula, calculate the three longest wavelengths in the Balmer series.

10. How much energy is required to remove an electron from a hydrogen atom in a state with $n = 8$?

11. A hydrogen atom is excited from a state with $n = 1$ to one with $n = 4$. (a) Calculate the energy that must be absorbed by the atom. (b) Calculate the display on an energy-level diagram the different photon energies that may be emitted if the atom returns to its $n = 1$ state. (c) Calculate the recoil speed of the hydrogen atom, assumed initially at rest, if it makes the transition from $n = 4$ to $n = 1$ in a single quantum jump.

12. Show on an energy-level diagram for hydrogen the quantum numbers corresponding to a transition in which the wavelength of the emitted photon is 1216 Å.

13. A hydrogen atom in a state having a binding energy (this is the energy required to remove an electron) of 0.85 ev makes a transition to a state with an excitation energy (this is the difference in energy between the state and the ground state) of 10.2 ev. (a) Find the energy of the emitted photon. (b) Show this transition on an energy-level diagram for hydrogen, labeling the appropriate quantum numbers.

14. What is the energy, momentum, and wavelength of a photon that is emitted by a hydrogen atom making a direct transition from an excited state with $n = 10$ to the ground state? Find the recoil speed of the hydrogen atom in this process.

15. From the energy level diagram for hydrogen, explain the observation that the frequency of the second Lyman series line is the sum of the frequencies of the first Lyman series line and the first Balmer series line. This is an example of the empirically discovered "Ritz combination principle." Use the diagram to find some other valid combinations.

16. An atom, *at rest* and in its ground state, is struck by another atom of the same kind which is also in its ground state but has a translational kinetic energy K. Show, from conservation principles, that the collision must be elastic if $K < 2E$, where E is the first excitation energy of the atom (i.e., if K is less than twice the energy from the ground state to the first allowed excited state in the atom's energy level scheme).

17. (*a*) Let ΔE be the energy difference between the normal and first excited state of an atom. Show that the kinetic energy K of an electron, mass m, must be at least

$$K = \left(1 + \frac{m}{M}\right)\Delta E$$

to excite an atom of mass M. (*b*) Compare the relation of K to ΔE for a Franck-Hertz experiment ($m \ll M$) to that for the situation in the previous problem ($m = M$).

18. (*a*) Show, using Coulomb's law and the dynamics of circular motion, that the frequency of revolution of an electron moving about a proton in a circle of radius r is given by $\nu^2 = e^2/(16\pi^3\varepsilon_0)mr^3$. (*b*) Show that the total energy E of this system, kinetic plus potential, is given by $E = -(e^2/8\pi\varepsilon_0 r)$. (*c*) Combine (*a*) and (*b*) to obtain Eq. 7-6,

$$\nu = \frac{4\varepsilon_0}{e^2}\sqrt{\frac{2}{m}}\,|E|^{3/2}.$$

19. (*a*) Prove that the dimension of the "Bohr radius," $a_0 = (4\pi\varepsilon_0)\hbar^2/me^2$, is that of length. (*b*) Prove that the "fine-structure" constant, $e^2/(4\pi\varepsilon_0)\hbar c$, entering the formula for orbital velocities, is dimensionless. (*c*) Prove that the dimension of the Rydberg constant, $R_\infty = me^4/8\varepsilon_0^2 h^3 c$, is that of inverse length.

20. In the ground state of the hydrogen atom, according to Bohr's model, what are (*a*) the quantum number, (*b*) the orbit radius, (*c*) the angular momentum, (*d*) the linear momentum, (*e*) the angular velocity, (*f*) the linear speed, (*g*) the force on the electron, (*h*) the acceleration of the electron, (*i*) the kinetic energy, (*j*) the potential energy, and (*k*) the total energy?

21. (*a*) Show that the smallest quantum number of the levels in hydrogen between which transitions giving rise to radio waves are possible is given by $n = \sqrt[3]{2R_H\lambda}$ where λ is the wavelength of the radio wave. (*b*) An important emission in radio astronomy is the 21-cm line from interstellar hydrogen. Find the corresponding value of n.

22. Prove the inequality of Example 5. Let $n \geq m + 1$.

23. Prove that Planck's constant has the dimensions of angular momentum.

24. Starting (in Bohr's model) from the classical mechanics of an electron moving about a nucleus in a circular orbit, show that applying the condition, Eq. 7-16, on the quantization of angular momentum leads directly to the Bohr stationary state energy condition, $E_n = -Z^2(me^4/8\varepsilon_0^2 h^2)(1/n^2)$. This is Eq. 7-13, which was derived earlier by means of the correspondence principle.

25. According to the correspondence principle, as $n \to \infty$ we expect classical results in the Bohr atom. Hence, the de Broglie wavelength associated with the electron (a quantum result) should get smaller compared to the radius of the orbit as n increases. Indeed, we expect $\lambda/r \to 0$ as $n \to \infty$. Is this the case?

★26. (*a*) Show that, when the recoil kinetic energy of the atom ($p^2/2M$) is taken into account, Eqs. 7-4 and 7-20 must be modified to

$$\nu \simeq \frac{\Delta E}{h} - \frac{(\Delta E)^2}{4mc^2 h}$$

and

$$\frac{1}{\lambda} = R_M Z^2\left(\frac{1}{m^2} - \frac{1}{n^2}\right) - \left(\frac{R_M^2 Z^4 hc}{4mc^2}\right)\left(\frac{1}{m^2} - \frac{1}{n^2}\right)^2$$

(*Hint.* Use the binomial expansion approximation and taken the recoil momentum to be $p = h\nu/c$). (*b*) Compare the wavelength of the light emitted from a hydrogen atom in the $3 \rightarrow 1$ transition when the recoil is taken into account to the wavelength without accounting for recoil. Express as $\Delta\lambda/\lambda$.

27. In the one-electron atom, the electron and nucleus revolve about their common center of mass with an angular velocity ω (see Fig. 7-11). (*a*) Show, from the definition of center of mass, that $R = (m/M)r$, where the mass and distance from the center of mass are m and r for the electron, and M and R for the nucleus, respectively. (*b*) Show that the quantization of total orbital angular momentum gives the relation $n\hbar = m\omega r^2(1 + m/M)$. (*c*) Show, if the nuclear charge is Ze, that the equation of motion can be written as

$$m\omega^2 r = \frac{1}{4\pi\varepsilon_0}\frac{Ze^2}{(R+r)^2} = \frac{1}{4\pi\varepsilon_0}\frac{Ze^2}{r^2}\left(\frac{M}{M+m}\right)^2.$$

(*d*) Then show, finally, that the quantized energies are given by

$$E_n = -\frac{\mu Z^2 e^4}{8\varepsilon_0^2 h^2}\frac{1}{n^2}.$$

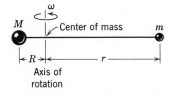

Figure 7-11. Problem 27.

28. What is the wavelength of the most energetic photon that can be emitted from a muonic atom with $Z = 1$?

29. In Chap. 5 we spoke of the positronium "atom," consisting of a positron and an electron revolving about their common center of mass, which lies halfway between them. (*a*) If such a system were a normal atom, show that the wavelengths of the emitted lines would be double that of the hydrogen atom (with infinitely heavy nucleus). (*b*) What would be the radius of the ground state orbit of positronium? (*c*) Assume that electron-positron annihilation takes place from the ground state of positronium. How, if at all, does this alter the γ-ray energies of the two-photon decay calculated in Chap. 5, where we ignored the bound system.

30. Using Bohr's theory, calculate the energy required to remove the electron from singly ionized helium.

31. Radiation from a helium ion He^+ is nearly equal in wavelength to the H_α line (the first line of the Balmer series). (*a*) Between what states (values of n) does the transition in the helium ion occur? (*b*) Is the wavelength greater or smaller than that of the H_α line? (*c*) Compute the wavelength difference.

32. In stars the Pickering series is found in the He^+ spectrum. It is emitted when the electron in He^+ jumps from higher levels into the level with $n = 4$. (*a*) State the exact formula for the wavelength of lines belonging to this series. (*b*) In what region of the spectrum is the series? (*c*) Find the wavelength of the series limit. (*d*) Find the ionization potential, if He^+ is in the ground state, in ev. Use $R_\infty = 109737$ cm^{-1}.

33. Assuming that an amount of hydrogen of mass number three (tritium) sufficient for spectroscopic examination can be put into a tube containing ordinary hydrogen, determine the separation of the first line of the Balmer series that should be observed. (Express as difference in wavelength).

34. A gas discharge tube contains H^1, H^2, He^3, He^4, Li^6, and Li^7 single-electron ions and atoms (the superscript is the atomic mass). (*a*) As the potential across the tube is raised from zero, which spectral line should appear first? (*b*) Give, in order of increasing frequency, the origin of the lines corresponding to the first line of the Lyman series of H^1.

35. Consider an electron in an atom of high atomic number whose binding to the nucleus is 1.0 kev and whose position is localized to within 1.0 Å. (*a*) Use the uncertainty principle to determine Δp. (*b*) Determine the momentum p and the fractional uncertainty $\Delta p/p$. (*c*) Comment on the applicability of the Bohr theory picture of electronic orbits to this case.

★36. Consider a particle of mass m and momentum p moving back and forth along the x axis between two points, $x = 0$ and $x = a$, from which it rebounds elastically. (*a*) Apply the Wilson-Sommerfeld quantization rules to find the predicted allowed values of the total energy E of the particle. (*b*) If $p \simeq \Delta p$, what is the lowest allowed value of E that satisfies the uncertainty principle?

★37. Consider a body rotating about an axis with angular momentum L. (*a*) Apply the Wilson-Sommerfeld quantization rules, and show that the possible values of the total energy are predicted to be

$$E_n = \hbar^2 n^2/2I, \qquad n = 0, 1, 2, 3, \ldots$$

where I is its rotational inertia (or moment of inertia) about the axis of rotation. (*b*) If $L \simeq \Delta L$, does $E = 0$ violate the uncertainty principle?

REFERENCES

1. E. Rutherford, "The Scattering of α and β Particles by Matter and the Structure of the Atom," *Phil. Mag.* **21,** 669 (1911). Reproduced in *Selected Readings in Physics—the Old Quantum Theory* by D. ter Haar, Pergamon Press, 1967.

2. Barbara Lovett Cline, *The Questioners,* Crowell Publishing Co., 1965, Chaps. 1 and 2.

3. E. N. de C Andrade, "The Birth of the Nuclear Atom," *Scientific American,* November 1956.

4. Leo Banet, "Balmer's Manuscripts and the Construction of His Series," *American Journal of Physics,* July 1970.

5. Neils Bohr, "On the Constitution of Atoms and Molecules," *Philosophical Magazine* **6,** 26 (1913). For an extract, see *Great Experiments in Physics,* edited by Morris H. Shamos, Holt-Dryden (1959).

6. E. H. S. Burhop, "Exotic Atoms," *Contemporary Physics,* **11,** No. 4, 1970.

7. Friedrick Hund, "Paths to Quantum Theory Historically Viewed," *Physics Today,* **19,** No. 8, August 1966, p. 23.

Appendix

Some Physical Constants

Avogadro's constant	N_0	6.02×10^{23} molecules/mole
Bohr radius	a_0	5.29×10^{-11} meter
Boltzmann's constant	k	1.38×10^{-23} joule/molecule K°
		8.63×10^{-5} ev/molecule K°
Electron charge to mass ratio	e/m_e	1.76×10^{11} coul/kg
Electron Compton wavelength	$\lambda_c = \dfrac{h}{m_e c}$	2.43×10^{-12} meter
Elementary charge	e	1.60×10^{-19} coul
Gravitational constant	G	6.67×10^{-11} nt-meter^2/kg^2
Permeability constant	μ_0	1.26×10^{-6} henry/meter
Permitivity constant	ϵ_0	8.85×10^{-12} farad/meter
Planck's constant	h	6.625×10^{-34} joule-sec
		4.135×10^{-15} ev-sec
Planck's constant/charge on electron	h/e	4.14×10^{-15} joule-sec/coul
Planck's constant times speed of light	hc	12400 ev-Å
		1.24×10^{-12} Mev-meter
Rydberg constant	R_∞	1.10×10^7/meter
Speed of light	c	3.00×10^8 meters/sec
		1.86×10^5 miles/sec
Stefan-Boltzman constant	σ	5.67×10^{-8} watt/meter^2 K°^{4}
Universal gas constant	R	8.31 joules/K° mole
Wien's displacement constant	b	2.90×10^{-3} meter-K°

Some Rest Masses

		kg	*amu*	*Mev/c^2*
Electron	m_e	9.1091×10^{-31}	0.000549	0.511
Proton	m_p	1.6725×10^{-27}	1.007277	938.256

Some Rest Masses (cont.)

Neutron	m_n	1.6748×10^{-27}	1.008665	939.550
Deuteron	m_d	3.3448×10^{-27}	2.013553	1876.43
Hydrogen atom	m_H	1.6734×10^{-27}	1.007825	938.777

μ^--meson $\qquad m_{\mu^-} = 207\, m_e$

π^+-meson $\qquad m_{\pi^+} = 273\, m_e$

Some Mass-Energy Conversion Factors

		Energy		Mass	
	Joule	ev	Mev	kg	amu
1 Joule =	1	6.242×10^{18}	6.242×10^{12}	1.113×10^{-17}	6.705×10^{9}
Energy 1 electron volt =	1.602×10^{-19}	1	10^{-6}	1.783×10^{-36}	1.074×10^{-9}
1 million electron volts =	1.602×10^{-13}	10^6	1	1.783×10^{-30}	1.074×10^{-3}
1 kilogram =	8.987×10^{16}	5.610×10^{35}	5.610×10^{29}	1	6.025×10^{26}
Mass 1 atomic mass unit =	1.492×10^{-10}	9.31×10^{8}	931.0	1.660×10^{-27}	1

Some Useful Constants, Numbers, and Relations

$$\hbar = \frac{h}{2\pi} = 1.054 \times 10^{-34} \text{ joule-sec} = 6.58 \times 10^{-16} \text{ ev-sec}$$

$$\frac{\hbar}{2} = \frac{h}{4\pi} = 0.527 \times 10^{-34} \text{ joule-sec} = 3.29 \times 10^{-16} \text{ ev-sec}$$

$\sqrt{2} = 1.414$	$\sqrt{3} = 1.732$	$\sqrt{5} = 2.236$	$\sqrt{10} = 3.162$
$\pi = 3.142$	$\pi^2 = 9.870$	$\sqrt{\pi} = 1.772$	$4\pi = 12.57$
$e = 2.72$	$1/e = 0.368$	$\log e = 0.434$	$\ln 2 = 0.693$

$\sin 30° = \cos 60° = 0.5000 \qquad \cot 30° = \tan 60° = 1.7321$

$\cos 30° = \sin 60° = 0.8660 \qquad \sin 45° = \cos 45° = 0.7071$

$\tan 30° = \cot 60° = 0.5774 \qquad \tan 45° = \cot 45° = 1.0000$

Mass-Energy relation $\qquad c^2 \qquad$ 931 Mev/amu

$\qquad\qquad\qquad\qquad\qquad\qquad\qquad$ 8.99×10^{16} joules/kg

Binomial Expansion

$$(x + y)^n = x^n + \frac{n}{1!}x^{n-1}y + \frac{n(n-1)}{2!}x^{n-2}y^2 + \cdots \quad (x^2 > y^2)$$

Some Conversion Factors

Mass

1 kg = 2.21 lb (mass) = 6.02×10^{26} amu = 5.610×10^{35} ev/c^2

1 slug = 32.2 lb (mass) = 14.6 kg

1 ton = 2000 lbs (mass) = 907.2 kg

Some Conversion Factors (cont.)

Length

> 1 meter = 39.4 in = 3.28 ft
> 1 mile = 1.61 km = 5280 ft; 1 in = 2.54 cm
> 1 meter = 10^{10} Å (angstrom units) = 10^{13} f (fermi)

Time

> 1 day = 86,400 sec
> 1 year = 365 days = 3.16×10^7 sec
> 1 sec = 10^6 μsec = 10^9 n sec

Angular Measure

> 1 rad = 57.3° = 0.159 rev
> 1 deg = 60 min = 3600 sec

Speed

> 1 mi/hr = 1.47 ft/sec = 0.447 meter/sec

Momentum

> 1 kg-meter/sec = 1.87×10^{27} ev/c = 1.87×10^{21} Mev/c

Electricity and Magnetism

> 1 amp = 1 coul/sec
> 1 weber/meter^2 = 1 tesla = 10^4 gauss

Force

> 1 nt = 10^5 dyne = 0.225 lb

Energy and Power

> 1 joule = 10^7 erg = 0.239 cal = 0.738 ft-lb
> 1 ev = 1.60×10^{-19} joule = 1.60×10^{-12} erg
> 1 horsepower = 746 watts = 550 ft-lb/sec

Answers to Problems

3. (a) $0.942c = 2.83 \times 10^8$ m/sec;
    (b) $621\ m_e$;
    (c) 212 Mev, 1.60×10^{-19} kg-m/sec.

4. (a) $1962\ m_e$;
    (b) $0.99999987c$;
    (c) $2.961\ m_e$, $0.940c$.

5. (a) $0.99881c$;
    (b) $0.108c$;
    (c) $20.5\ m_e$, $1.01\ m_p$.

9. (a) 2.56×10^5 volts;
    (b) $0.745c$;
    (c) 1.365×10^{-30} kg, 4.09×10^{-14} joules.

14. (a) 0.145 weber/m^2;
    (b) 1.98.

15. 663 km.

17. 1.82×10^{20} m, 1.22×10^9.

18. $268\ m_e$, π meson.

19. 114.2 km above sea level.

20. 4.42×10^{-36} kg, 2.208×10^{-32} kg.

21. 4.47×10^{-6} nt/m^2.

22. 5.62×10^{26} Mev.

23. 4.22×10^{-12} kg, $4.66 \times 10^{-13}\%$.

24. (b) 0.511 Mev;
    (c) 938 Mev.

25. (a) 2.7×10^{14} joules;
    (b) 1.79×10^7 kg;
    (c) 5.96×10^6.

26. 92.1 Mev.

27. $2.5\ m_0$.

28. (a) $\frac{c}{3}$;
    (b) $2.12\ m_0$.

29. (a) $(7/12)\ m_0 c$;
    (b) $(1/5)\ m_0 c$;
    (c) $(35/12)\ m_0 c^2$;
    (d) $(34.29/12)\ m_0$;
    (e) $(0.71/12)\ m_0 c^2$.

32. (a) 2.36×10^{-21} kg-m/sec;
    (b) 8.78×10^{-4} Mev.

34. 29.6 Mev (neutrino), 4.15 Mev (muon).

35. (a) 8000 m;
    (b) $0.884c$;

    (d) $\theta' = \tan^{-1}\left[\dfrac{4\sin\theta}{5\cos\theta - 3}\right]$;

    (e) 2×10^4 cps.

Chapter 4

1. 4830 Å.

2. (a) 4.1×10^9 kg/sec;
    (b) 6.45×10^{-14}.

3. 4000°K,
    5500°K,
    7000°K.

4. 9.35×10^{-6} m.

5. 5460 Å.

6. (a) 2.898 Å;
    (b) 4.28×10^{-3} Mev.

7. (b) 282°K.

8. White or slightly bluish.

9. Betelgeux ~ 4000°K,
    Sun ~ 6000°K,
    Rigel > 7000°K.

12. (a) 7.53 watts;
    (b) 2.09 photons/sec.

15. (a) 0.04 ev;
    (b) 1.54×10^4°K.

18. 1.62×10^4°K.

19. (a) ΔE: 10.21 ev,
              12.10 ev,
              1.89 ev;
    (b) ν: 2.46×10^{15} cps,
           2.92×10^{15} cps,
           4.56×10^{14} cps;
    (c) λ: 1220 Å,
           1030 Å,
           658 Å.

20. 6.63×10^{-34} joule-sec.

Chapter 5

1. (a) No;
    (b) 5400 Å.

2. (a) 2.0 ev;
    (b) Zero;
    (c) 2.0 volts;
    (d) 2950 Å.

4. 3820 Å.

5. (a) 6.59×10^{-34} joule-sec;
    (b) 2.28 ev;
    (c) 5440 Å.

6. Lithium, Barium.

7. (a) 1.99×10^{-6} m;
    (b) 10^{-17} sec.

8. 0.48 Å.

9. (a) 3120 ev;
    (b) 1.44×10^4 ev.

10. 1.232×10^{20} sec^{-1},
    0.0244 Å,
    2.72×10^{-22} kg-m/sec^2.

11. (a) 5.90×10^{-6} ev;
    (b) 2.04 ev.

12. 3.72×10^{21} photons/cm^2-sec.

13. 3.61×10^{-17} watts.

14. (a) 887 m;
    (b) 1.97×10^6 photons/cm^3.

21. 0.682 Mev.

22. $K = 2h^2\nu^2/(m_0 c^2 + 2h\nu)$

23. (a) 0.027 Å, 0.060 Mev;
    (b) 0.060 Å, 0.309 Mev.

24. $p_x = 5.33 \times 10^{-22}$ kg-m/sec,
    $p_y = 4.48 \times 10^{-23}$ kg-m/sec.

26. 2.64×10^{-5} Å.

27. (b) 0.0667 Å.

28. (a) 511 Kev, 12.4 Kev;
(b) 1020 Kev.

29. 1876 Mev.

32. (a) 2.02 Mev;
(b) 29.6%.

33. (a) 5.44×10^{-22} kg-m/sec;
(b) 2.69 ev, Yes.

35. (a) 0.51 Mev.

Chapter 6

1. (a) 1.656×10^{-35} m.

2. 4.34×10^{-6} ev.

3. (a) 6.21×10^{-3} Mev/c,
6.21×10^{-3} Mev/c;
(b) 0.51 Mev, 6.21 Kev;
(c) 6.05×10^{-3}.

4. (a) 3.88×10^{-2} ev;
(b) 1.45 Å.

5. 8.78×10^{-3} Å, 1.24×10^{-2} Å,
2.86×10^{-4} Å.

6. (b) 10,200 volts.

7. (a) 20.4 Kev;
(b) 37.6 Kev.

8. 2.48×10^{-17} m $\simeq r_0/60$.

10. (b) 59.4°;
(c) 182 volts at 90°

11. $\theta_{ph}/\theta_e = 5.12$.

13. (a) 41.3°;
(b) 1.14 Å.

14. 39.4 KV.

15. (a) 15.13 Kev;
(b) 0.124 Mev, Gamma rays;
(c) Electron.

17. (a) 398 Å;
(b) 398 microns,
(c) 0.398 m.

19. $\Delta p_y/p = 7$ parts/1 billion.

23. 2.89×10^6 m/sec.

24. (a) 5.27×10^{-25} kg-m/sec, Yes;
(b) 5.27×10^{-21} kg-m/sec, No;
(c) 5.27×10^{-21} kg-m/sec, Yes.

25. 3.3×10^{-4} ev.

26. 1.33×10^{-7} sec.

Chapter 7

3. 2.95 Å.

6. 1.58×10^{-14} m.

7. 32.6 Mev.

8. (a) 0.0195 K;
(b) 2.754×10^{-14} m.

9. 6560 Å, 4860 Å, 4340 Å.

10. 0.213 ev.

11. (a) 12.75 ev;
(b) 0.66 ev, 1.89 ev, 2.55 ev, 10.20 ev,
12.09 ev, 12.75 ev;
(c) 4.065 m/sec.

13. (a) 2.55 ev.

14. 13.46 ev, 7.18×10^{-27} kg-m/sec, 923 Å,
4.29 m/sec.

20. (a) $n = 1$;
(b) 0.529 Å;
(c) 1.054×10^{-34} kg-m²/sec;
(d) 1.99×10^{-24} kg-m/sec;
(e) 4.14×10^{16} sec^{-1};
(f) 2.19×10^6 m/sec;
(g) 8.25×10^{-8} nt;
(h) 9.07×10^{22} m/sec²;
(i) 13.6 ev;
(j) -27.3 ev;
(k) -13.6 ev.

21. 1660.

26. (b) 3.25×10^{-9}.

28. 4.91 Å.

29. (b) 1.058 Å.

30. 54.4 ev.

31. (a) $n = 6$ to $n = 4$;
(b) Smaller;
(c) 2.68 Å.

32. (a) $\dfrac{1}{\lambda} = 4R_{\mathrm{He}}\left(\dfrac{1}{16} - \dfrac{1}{n^2}\right)$;
(b) near infrared;
(c) 3646 Å;
(d) 228 Å.

33. 2.38 Å.

34. (a) K_α of H^1;
(b) H^1, H^2, He^3, He^4, Li^6, Li^7.

35. (a) 5.27×10^{-25} kg-m/sec;
(b) 1.71×10^{-23} kg-m/sec,
3.08×10^{-2}.

36. (a) $\dfrac{n^2h^2}{8ma^2}$;
(b) $\dfrac{h^2}{8ma^2}$.

Index